CONSIDER A SPHERICAL PATENT

CONSIDER
A SPHERICAL
PATENT

IP and Patenting in
Technology Business

Joseph E. Gortych

CRC Press
Taylor & Francis Group
Boca Raton   London   New York

CRC Press is an imprint of the
Taylor & Francis Group, an **Informa** business

CRC Press
Taylor & Francis Group
6000 Broken Sound Parkway NW, Suite 300
Boca Raton, FL 33487-2742

© 2014 by Taylor & Francis Group, LLC
CRC Press is an imprint of Taylor & Francis Group, an Informa business

No claim to original U.S. Government works

Printed on acid-free paper
Version Date: 20130930

International Standard Book Number-13: 978-1-4398-8805-6 (Paperback)

Library of Congress Cataloging-in-Publication Data

Gortych, Joseph E.
Consider a spherical patent : IP and Patenting in Technology Business / by Joseph E. Gortych.
pages cm
Includes bibliographical references and index.
ISBN 978-1-4398-8805-6 (alk. paper)
1. Patents. 2. Technological innovations--Patents. 3. Intellectual property. 4. Inventions. I. Title.

T211.G67 2014
608--dc23   2013036571

Visit the Taylor & Francis Web site at
http://www.taylorandfrancis.com

and the CRC Press Web site at
http://www.crcpress.com

To Jennifer and Sam

CONTENTS

Contents

PREFACE

A mentor and colleague, the late Doug Goodman, once said to me, "There are no boring subjects—only boring books and boring people." Some people think that the patent field generates boring books and attracts boring people. That might have been the case 50 years ago, when information about patents was relegated to arcane legal texts and the occasional back-page news story of interest mainly to a nerdy subculture of lawyers called "patent attorneys."

Not anymore. The patent field is now at the forefront of both business and legal interests. A variety of nonlawyer professionals are now heavily involved in patents and patenting alongside nerdy and non-nerdy patent lawyers. There are now plenty of nonlegal books about patents. Patents are also showing up in the news. Hardly a week goes by without a front-page or main web page news report of a matter concerning someone's "intellectual property"—the somewhat pretentious catch-all phrase (typically just called "IP") for copyrights, trademarks, patents, trade secrets, ideas, and know-how. Sometimes the story is a major patent infringement lawsuit with billions of dollars at stake or the purchase of a gargantuan patent portfolio by a company seeking to level the playing field in a tight but lucrative market. Sometimes it is smaller—like the news item of today's writing, which is that Frito Lay lost a lawsuit against Ralcorp, alleging unsuccessfully that the latter stole its IP relating to scoop-shaped tortilla chips. At stake was $4.5 million—a relatively paltry sum in IP terms.

The Supreme Court of the United States has also been taking on more cases relating to IP in general and patents in particular. This is a clear sign of the increasingly important role that patents and other forms of IP play in today's knowledge-based economy. In news recent to this writing, the Supreme Court ruled that naturally occurring DNA sequences are "products of nature" and therefore not patentable.

It is as though patents recently completed a two-century metamorphosis from staid legal documents to dynamic business tools. This metamorphosis has helped usher in what I call the *IP Era of the Information Age.* Small and large businesses alike, as well as academic institutions and independent inventors, are now leveraging and monetizing patents in increasingly creative and assertive (and in some cases, downright obnoxious) ways. This manufacturing and leveraging of ideas is now as commonplace as the manufacturing and selling of products and, when done properly, can be far more lucrative.

Those ideas, innovations, creations, inventions—call them what you will—do not just appear out of nowhere. They are actively generated from the minds of all levels of people working in technology businesses. As companies increasingly come to rely on IP for their competitive edge, more and more people involved in technology business are being nudged if not shoved into actively participating in their company's or organization's IP activities. The fact is that there are fewer and fewer places for people to hide and avoid dealing with IP and patents in particular. In fact, I'd say it's already too late. If you work at a technology business, you are already working with innovations, inventions, and patents, even if you don't know it yet.

The prominence of patents in today's economy has stirred among technology businesses a great deal of hype and happy talk about wanting to be "leaders in innovation" and to "leverage" and "monetize" their IP. Yet, too many of the people who are supposed to be doing the innovating—the technicians, engineers, scientists, designers, programmers, and every other kind of technology worker who already has way too much to do—don't know much, if anything, about patents. Many don't understand their own company's thinking about patents and the company's own internal process for obtaining patents for its inventions.

While most technology businesses are filled with technology workers fully capable of inventing great inventions, it is not unusual for only a small percentage of those workers to be involved actively in the company's patenting system. In the meantime, too many managers and executives are disengaged from their organizations' patenting processes. It is easier for them to throw all the IP over the wall to a law firm, just to have one less thing to occupy their already full plates.

The result is that, while our economy has become IP centric, our technology workforce has remained largely IP illiterate. I can speak with some authority on this subject. In the early 1990s, I could have been the poster child for a campaign to "End IP Illiteracy Now." When I was first hired as an associate engineer at IBM, I had absolutely no idea what a patent was. At no time during my undergraduate physics education and my graduate engineering education were patents discussed. When I joined IBM, I was a full-fledged member of the *IP illiterati*. And trust me, I wasn't alone.

Going from being clueless about patents to working for a business that viewed inventing and patenting as fundamental parts of an engineer's job was a total mind-bender. It took me a long time and a lot of work to get over my ignorance about patents and gain some inkling of what patents were and how they were used in business. Eventually, I got comfortable enough with the innovating part and ended up being named as an inventor on a few IBM patents and serving on an invention review board. However, along the way, there was a lot to learn as well as a fair amount of patent nonsense and misinformation to parry.

I wrote this book for two main reasons. First, I know that there are hordes of technology workers out there who have been asked to contribute to their company's IP effort but simply cannot. In some cases, they don't know enough about patents even to know where to begin. In other cases, their company has such a dysfunctional patenting system that the task is nearly impossible.

Second, while there are plenty of books on patents, for whatever reason, not enough technology workers are reading them. While I make no claim here that this work is the be-all and end-all book on patents, I try to provide some key insights into the modern patenting scene from different perspectives and viewpoints that "technology workers" will hopefully find engaging. I've tried to write the kind of book that I wish had been available when I was learning about patents as part of my job as an engineer.

Here is an additional reason to read this book and others like it

If you work at a technology business and choose to remain aloof about patents and the patenting process, it might one day cost you your job.

The same can be said for a technology business that remains aloof about patents: It may one day cost the business its very existence.

People who work at a technology business need to understand some fundamentals about patents, the ins and outs of how they are obtained, how they are used in business, the various reasons why they are pursued, and how to participate effectively in their organization's patenting process. Like it or not, the days when it was optional to know something about patents are officially over. The IP Era of the Information Age demands a new IP professionalism. It requires people to be informed about patents and, more generally, informed about their company's IP and how it is used and leveraged. Those who choose to remain ignorant about IP matters in general and patents in particular are choosing to forfeit a key technical skill at their peril. On the flipside, those who choose to become IP literate and to participate actively in their company's patenting system are going to stand out and be much more employable over the long run.

PRELIMINARY REMARKS

A few remarks are needed before we get to an overview of the contents of this book:

- I include at the end a glossary of useful terms and abbreviations that are used in the book.
- One term I use often in the following pages is "patenting system" and it refers to the internal processes and procedures a technology business uses to obtain patents. The United States Patent and Trademark Office (USPTO) is the government body that manages the US patenting system for the United States at large, and it falls under the Department of Commerce.
- There is a Dilbertian tendency in business to see the world in terms of "managers" and "nonmanagers." Indeed, some of the topics in this book speak directly to nonmanagers, while others speak directly to managers. This moving between audiences is intentional and is part of viewing IP and patents from different perspectives. Hopefully, it will illustrate some of the potential sources of polarization within the patenting system of a technology business. If you consider yourself a nonmanager, then you should know about the problems that managers face when it comes to IP and vice versa.
- Parts of this book take a critical look at the USPTO and the examination process. Nothing about this critical look is meant to offend those who work at the USPTO, particularly the patent examiners. Virtually everyone I have dealt with at the USPTO has been extremely helpful, courteous, knowledgeable, and professional. The comments in this book about

the USPTO are directed mainly to its inherently bureaucratic nature and its effect on the patents that issue from it. As I try to remind those who complain about the USPTO, those who have to work in a bureaucracy the size of the USPTO are at its mercy and have little control over the rules it imposes on them.

- No book related to legal matters would be complete without the following caveat: The information presented herein is not intended as legal advice. As you will soon appreciate, patenting issues can be complex and nonlinear. If you have a patent-related issue, you should seek the help of a qualified patent attorney.
- The content of this book is bound up with my opinions, which are not those of my clients, colleagues, people who have reviewed this book, or anyone else. Reasonable people may disagree with the perspectives and viewpoints expressed in this book.
- I welcome civilized and thoughtful comments about what you read in these pages. Feel free to send me an e-mail at sphericalpatent@opticus-ip.com

ORGANIZATION AND OVERVIEW

To be approachable and readable, this book is by necessity incomplete. There are far too many patent-related topics to cover in a single book. I have endeavored to address the topics that I believe are most relevant to someone working in or otherwise involved with a technology business.

Chapter 1, "Overview and Underview," introduces the metaphor of the "spherical patent" as it relates to taking a simplified (if not oversimplified) view of patents. We then define the key terms "technology worker" and "technology business" and position these players within the larger world of patent-related services. Finally, we learn about the main interconnections between IP and the typical functions of a technology business and the main IP-related interactions a technology business has with the outside world.

Chapter 2, "The IP Universe," opens with a bang, so to speak, by explaining how the IP universe began and then evolved through the expansion of the byzantine array of laws, regulations, rules, and procedures that govern patents. We also introduce the concept of business entropy as a measure of the amount of disorder in a business system—in our case, a technology business's internal patenting system. We end with an overview of the common forms of IP misinformation that pervade many a technology workplace and that contribute to IP system entropy.

Chapter 3, "Beyond the Spherical Patent," explains the numerous "requirements" that a patent application must satisfy to issue as a patent. We observe that the patent office does not actually *measure* any of these requirements but rather *filters* patent applications based on the requirements. We see how some of the requirements do not fall within a patent examiner's natural field of view. We also consider the irony that *business value,* which is arguably the most important feature of a patent, is *not* one of the requirements.

Chapter 4, "Patent Portfolio Thermodynamics," explores how patents can be viewed in groups called *patent portfolios*. We see that when a patent portfolio gets large enough, it exhibits characteristics that can be described by "patent thermodynamics." These dynamics are akin to the thermodynamics used to explain the behavior of gasses.

Chapter 5, "Classical and Quantum Patent Mechanics," opens with a discussion of how "classical patent mechanics" allows us to view patents based on the certainty of outcomes. We then learn that this viewpoint has serious shortcomings and that patents are better viewed using "quantum patent mechanics," which accounts for the underlying uncertainty that pervades the patenting process.

Chapter 6, "Provisional Patent Applications Revealed," takes a critical look at the much used and often abused provisional patent application. We see how drafting a provisional application based on spherical patent assumptions can create serious problems down the road should any follow-on patent or patents gain the serious attention of a competitor.

Chapter 7, "The Double-Edged Sword of Infringeability and Validity," explores the interrelation between patent "infringeability" and patent validity. We discuss "zones" along the infringeability/validity continuum according to which claims can be loosely categorized. The frustrating phenomenon whereby the patent office issues patents of dubious validity is explored in the contexts of classical and quantum patent mechanics.

Chapter 8, "Lost in IP Space," examines the exponential growth in patenting through the concept of the *fractal nature of innovation*. We then discuss the nature of *IP space*, including how and where patents reside in the space and how their status as prior art affects new patents entering the space. We also discuss the timing of when patents issue in relation to the development and commercialization of a technology. This chapter also covers the topics of prior-art searching and freedom-to-operate analysis.

Chapter 9, "Patent System Operational Reality," explores the real-life difficulties of getting a patenting system of a technology business to function properly. We describe the common elements of a best-practice patenting system for a technology business, including the underappreciated importance of properly documenting innovations. We discuss the innovation review process and the importance of balancing the dynamic interplay between the business, legal, and technical aspects of a company's patenting system.

Chapter 10, "That's Obviousness!" delves into the important role that obviousness plays in both the patenting system of a technology business and the patent examination process at the USPTO. We provide a history of obviousness as a legal concept to gain some perspective on how the courts, as the ultimate measurers of obviousness, have viewed the subject in some key patent cases. We also review the USPTO's examination guidelines for analyzing obviousness to see firsthand what the standards for assessing obviousness are supposed to be.

Chapter 11, "Inventions and Inventors," begins by discussing the inconvenient problem of IP illiteracy among the technology workforce. A case is made for a new IP professionalism that is appropriate for the IP Era of the Information Age. The importance of proper IP training and education, including mentoring programs, is emphasized. We reflect upon the benefits of technology workers achieving a state of IP *zanshin*. We then take a critical look at the pros and cons of invention reward/award systems.

Chapter 12, "Independent Inventors," addresses the special patent-related challenges that independent inventors face. I offer the "Independent Inventor Riot Act," which every would-be independent inventor should read and reflect upon before starting to write checks if the inventor is to rid himself or herself of illusions and delusions about how easy it is to make money from patents.

Chapter 13, "Central Organizing Principles and Patent Strategies," uses the concept of a "central organizing principle" (COP) to articulate what an IP strategy is and how to formulate one. We review various basic COPs and the corresponding IP strategies, and examine how some technology businesses utilize those COPs and execute the corresponding IP strategies in practice.

ACKNOWLEDGMENTS

The author would like to thank the following people who reviewed either some or all of this book: Dr. Paul Lett, Dr. Ekaterina Rogacheva, Thomas Leach, Dr. Brian Caldwell, Gary Russell, Larry Meier, and Keith Roberson. Whatever shortcomings this book may have, they would have been far worse if not for these reviewers.

Heartfelt thanks go out to Penelope Cray, who did a terrific job editing the text and formatting the chapters. Many thanks to Dr. Robert Lawson Brown for generating the Apollonian gasket diagram in Figure 8.4 of Chapter 8. Thanks also to Wolfram Research for permission to use their Apollonian gasket sequence for Figures 8.3(a) through 8.3(f) in Chapter 8. Jonathan Chang created the yin–yang graphic of Figure 9.3 in Chapter 9.

I'd also like to thank Dr. Nat Ceglio for his insights into the use of "central organizing principles" for everything in life that matters, including patent strategies. I'm also grateful to the following people who helped me learn about patents and the patenting process when I worked at IBM as an optical engineer: John Cronin, Mark Chadurjian, William Skladony, Rick Kotulak, Larry Meier, Mike Hibbs, Peter Mitchell, and Alan Rosenbluth.

I'm grateful to all of the excellent professors and mentors I have had along the way in my education, especially including at Rutgers University: Dr. Mohan Kalelkar, Dr. Marc Croft, Dr. Gordon Thompson, and Dr. Michael Beals; and at the University of Rochester: Dr. Dennis G. Hall and Dr. Duncan Moore, as well as then-colleagues Dr. Susan Houde-Walter, Dr. Laura-Weller Brophy, Dr. Bülent Kurdi, and Dr. Martijn deSterke. At Vermont Law School, Professor Oliver Goodenough was very generous with his time in facilitating an independent study of patents and intellectual property.

Finally, many thanks to my clients, past and present, for their trust and confidence, and to the many inventors with whom I have had the privilege to work.

ABOUT THE AUTHOR

Joseph E. Gortych (jg@opticus-ip.com) is president of Opticus IP Law PLLC, an intellectual property (IP) law firm based in Sarasota, Florida, and specializing in optics, photonics, and semiconductor technologies (www.opticus-ip.com). His practice emphasizes the strategic and common-sense use of patents in business.

He received his BS in physics from Rutgers University, his MS in optics from the University of Rochester's Institute of Optics, and his JD from Vermont Law School. Mr. Gortych is a registered patent attorney and is an active member of SPIE (the international society for optics and photonics) and the Optical Society of America.

Mr. Gortych writes, lectures, and provides legal counsel on patent-related IP matters for technology businesses in the United States and abroad. He also consults with technology businesses and organizations on structuring and operating their internal patenting systems. He is a named inventor on eight US patents. In his spare time, he enjoys sailing and training in Brazilian jiu-jitsu.

CHAPTER 1

OVERVIEW AND UNDERVIEW

1.1 SIMPLIFICATION VERSUS OVERSIMPLIFICATION

There is a joke that is well known in the physics community, and it goes something like this:

> A dairy farmer asks a physicist friend to solve a problem the farmer is having at his dairy farm. The physicist investigates the dairy farm and then goes away. He returns a week later, and the farmer asks the physicist if he has come up with a solution to the problem. The physicist says, "Yes, I have an excellent solution, but it applies only to a spherical cow in a vacuum."

This joke reflects the propensity of (and indeed the need for) physicists to make simplifying assumptions about complex phenomena in order to make their analysis tractable. Such simplifying assumptions can lead to great insights. For example, in 1939 the physicists J. Robert Oppenheimer and Hartland Snyder gleaned the existence of black holes by making some rather extreme simplifying assumptions about an imploding star, not the least of which was that the star had a perfectly spherical shape (Thorne 1994).[1] Likewise, the assumption that atoms are tiny solid spheres works well for understanding the basic thermodynamics of a gas. String theory, which posits that all matter is fundamentally made of different manifestations of mind-bendingly small energy strings, has progressed far on the assumption that all extra dimensions beyond the familiar three spatial dimensions and the one time dimension have circular- and sphere-based shapes (e.g., hyperspheres). More recently in the news, the Chelyabinsk meteor that exploded over Russia on February 15, 2013, was said to have been an asteroid with a mass of about 10,000 metric tons and a diameter of about 20 meters, *assuming a spherical asteroid.*

The flip side of making simplifying assumptions about a complex phenomenon is that it can obscure insight into other important aspects of that phenomenon. For example, the simplifying assumptions Oppenheimer and Snyder made precluded them from understanding what happens inside a black hole. The assumption that atoms are tiny, solid spheres utterly fails when we try to explain certain behavior of atoms in a gas, such as their ability to form molecules or be radioactive. The assumption in string theory that the extra dimensions are circular or even hyperspherical doesn't quite yield all the correct answers and indicates that the actual multidimensional shapes are much more sophisticated. And most photos of asteroids seem to show that they generally have weird shapes that are anything but spherical.

1.2 THE SPHERICAL PATENT

I use the term "spherical patent" to refer to an oversimplified view of a patent or patents and patenting in general. A sphere, after all, looks the same from all directions. It is arguably the purest geometrical form. It has no distinguishing features other than its perfect symmetry. So the term "spherical patent" reflects the conscious or unconscious choice people (including the author) make when regarding all patents as simple and all the same. It reflects the conscious and unconscious simplifying assumptions and approximations that people make about patents while ignoring their internal structure, interactive properties, requirements for getting issued, why they are pursued, and how they are used and work in practice. In places in this book, we need to assume a spherical patent to simplify the discussion. The spherical patent assumption provides insight at the expense of accuracy.

Too many technology businesses and the people that work in them are content to always view patents as spherical. Rather than develop some institutional patent savvy and sophistication by getting into the details, some simply abdicate responsibility and turn everything over to a law firm. This approach is OK when the details being delegated are strictly legal in nature. But patents aren't just *legal* documents. They are also *business* documents and *technical* documents. So going beyond a spherical patent view to understand the complexity and multidimensional nature of patents will allow a technology business to better appreciate what it needs to be doing internally and what responsibility it owns with its patents, as well as which aspects and activities it can delegate or outsource.

1.3 THE COMMON WORKER DRIVES INNOVATION

At their very core, patents are used to protect certain kinds of innovations called "inventions." They provide a time-limited monopoly on the invention (20 years, assuming a spherical patent) in exchange for the patent owner disclosing the invention in sufficient detail.

Where do patentable inventions come from? Adam Smith noted in his seminal work *An Inquiry into the Nature and Causes of the Wealth of Nations,* originally published in 1776, that "a great part of the machines made use of in those manufactures in which labor is subdivided, were originally the inventions of common workmen" (Smith 2000). The same is true in today's Information Age, where, despite air-conditioned offices and computers, many of us (including the author), however begrudgingly, fall into Smith's category of "common workers."

In this book, I group all who work in a technology-based business or organization, whatever their jobs, under the common rubric "technology worker." Anyone who has to innovate as part of his or her job is a technology worker. Anyone who works with or interacts with innovations is a technology worker. The most common types of technology workers are technicians, engineers, and scientists. But my job as a patent attorney working with cutting-edge inventions makes me a technology worker as well. Computer programmers are technology workers. People who manage others involved directly with technology are technology workers.

I use the term "technology business" to refer to any business or organization that is involved with technology and that relies on some level of innovation to improve its technology. This term thus encompasses high-tech companies, medium-tech companies, and even some low-tech companies, as well as academic institutions that seek to generate and commercialize inventions through their technology transfer offices. Patent law firms are technology businesses. Consulting firms that work with technology are technology businesses. Independent inventors are technology businesses to the extent they are trying to leverage what they invent for profit.

While concepts described in this book may apply to some degree to other non-technology businesses as well, such as railroads, parts-supply companies, novelty companies, certain kinds of independent inventors, and others that may be dabbling in patents, this book is directed more toward businesses and organizations that are serious about patenting and that need to use patents as business tools for improving or ensuring their economic livelihood.

The technology workers of the IP Era of the Information Age are much like the blue-collar workers of yesterday's Industrial Age in that their importance in driving today's complex, technology-based economy cannot be overstated. Every time a politician mounts a soap box and harangues about how America needs to "increase innovation to stay competitive in the world markets" and "boost high-tech manufacturing," he or she is speaking directly to and about technology workers. Technology workers are the ones shoveling the coal into the economic engine. They produce the threads of innovation that are woven into the cloth of patents and that make up the never-ending line of "new and improved" technology products for which the world has a seemingly insatiable appetite.

1.4 THE ADVERSE IMPACT OF ÜBERPRODUCTIVITY ON INNOVATION

To drive the technology economy at an ever faster pace, technology businesses are always looking for inventive ways to increase technology worker productivity. While the do-more-with-less business mentality has improved many corporate bottom lines, it also has exacted a toll on the health and sanity of many technology workers. Ironically, the technologies invented and improved by technology workers often render many of these workers obsolete or make them work harder and longer. Those that remain employed must work asymptotically closer to their theoretical maximum working capacity in increasingly smaller work spaces.

This drive toward überproductivity has numerous unintended consequences, not the least of which is the impairment of the technology worker's ability to innovate and actively participate in a patenting system as part of his or her job. It also wreaks havoc on the patenting system of a technology business.

This sentiment can be summarized in the following axiomatic but underappreciated statement:

> **No Time to Think = No Innovation**

1.5 THE EVOLUTION (AND DEVOLUTION) OF PATENTING IN TODAY'S ECONOMY

At the beginning of the twentieth century, the world's more advanced economies were driven by the innovations of the Industrial Revolution and the aforementioned blue-collar workers of yore. Back then, businesses made money the old-fashioned way: by manufacturing and selling products. The economies of this new century are more unified and flat, and technology businesses are making money the new-fashioned way: by manufacturing and selling "ideas" and, in particular, inventions.

This economic transition from "sweat of the brow" to "sweat of the brain" spurred the evolution of IP, which sparked the economic revolution that brought us into the Information Age. The predominant role of IP in the world economy has driven us into the IP Era of the Information Age.

The Information Age's increased economic and technological complexity has translated into increased patenting complexity. Much of the technology developed and patented in the Information Age has made it possible to invent things faster and to file patent applications more frequently, which in turn allow us to invent things even faster and file patent applications even more frequently. It is little wonder, then, that the plow-horse US patenting system instituted in the late eighteenth century struggles to keep pace with the racehorse rate at which technology businesses innovate to capture ever greater market shares and squeeze out additional profits.

This increased technological complexity also means that patents themselves have become on average more complex. Where the patents of 100 years ago were typically a few pages long and had a few figures and maybe a dozen claims, the patents of today commonly run thirty pages and have fifty or more claims and dozens of figures. This increase in patent complexity has in turn increased the burden on the United States Patent and Trademark Office (USPTO) to process patent applications. The types of subject matter that can be patented have also recently been expanded to include software and business processes. This expansion has fueled a mini gold rush of sorts, as companies race to lay claim to and squeeze value from this newly available IP real estate. This has served only to further strain the USPTO's limited resources.

The change in the nature of patenting over the last 25 years or so resembles the change in the US health-care system. Fifty years ago, doctors were the primary providers and managers of health care. Somewhere along the line, businesspeople—in particular, insurance companies—figured out that they could make serious money by getting a piece of the health-care action. Fast forward to today, where insurance companies are the main players. They drive the health-care system and have a heavy hand in setting the rules. Doctors now take orders from the insurance companies. The transition also generated new kinds of jobs and transformed existing jobs. Huge numbers of managers, data-entry personnel, information technology (IT) staff, and IT systems are needed just to deal with the myriad arcane health-care regulations and paperwork.

Similarly, 50 years ago, lawyers were the primary providers and managers of patents and IP. Somewhere along the way, businesspeople realized that there was serious money to be made in taking control of and becoming the drivers of IP. There are now more business-related IP jobs than ever before, ranging from the IP liaisons that work with inventors and patent attorneys and patent agents to IP managers, chief IP officers, nonattorney IP consultants, and licensing professionals. While this transition to a more business-driven IP industry has not been as comprehensive as it has been for the health-care system, the shift in control from the IP lawyers to nonlawyer businesspeople is profound and irreversible.

1.6 THE IP BAZAAR

The above described tectonic shift in the IP landscape dramatically changed the cast of characters involved in patenting and IP-related pursuits. What was once essentially the exclusive purview of staid patent law firms and the occasional invention promoter has evolved (or devolved, as the case may be) into a veritable IP bazaar. IP consulting companies have sprouted up, offering to help assist and manage a company's internal IP processes and get it better patents. Some patent law firms, while at first a bit slow on the uptake, have smartened up and now offer remarkably similar kinds of IP consulting services and talk more than ever about "IP strategy" and "IP value."

In the meantime, while some of the dubious invention promoters are still hanging around, hip consulting companies have sprung up that offer to "monetize" (the de rigueur word for "market and sell") others' patents for a piece of the action.[2] If you're short on cash and have a great invention, no worries. You can get "patent pending" status today online for as little as $149. Intellectual property searching companies offer sophisticated Internet-based search tools for finding the most obscurely published prior art. Companies have set up websites that allow for searching and downloading of patents for free.

Some outfits post want ads for *prior art* (that is, references, information, etc., that predate a given invention) and will pay from a few hundred dollars to several thousand for good submissions. IP analysis companies offer to analyze patent databases using sophisticated software and create impressive looking patent maps of an IP space, and to process patent-related data in ways limited only by the imagination and computing power. Patent-outsourcing firms in exotic lands send mysterious e-mails (not unlike those from that bothersome nephew of that wealthy Nigerian prince) offering to save you money by drafting your patent applications and performing other kinds of legal work from afar.

For those who like shopping on the Internet, there are online patent exchanges that sell patents, typically with a "no return or refunds" policy. If the excitement of live bidding is more appealing, there are IP auctions where you can buy up IP real estate for pennies on the dollar, or for dollars on the penny, as the case may be. There are now technology companies whose entire business models are based solely on patenting inventions and relying almost exclusively on patent licensing income and the not-so-subtle threat of litigation to encourage licensing deals. There are also companies called *nonpracticing entities* (NPEs) or, more pejoratively, "patent trolls" that obtain patents for the sole purpose of inviting (and if the invitation is declined, compelling) royalty payments.

To counter the possible adverse effects of NPEs, patent collectives have sprung up that buy patents and sell access to the collection ("portfolio") to prevent others (namely, the NPEs) from obtaining and trying to enforce the patents (presumably it costs less to join the collective than to pay off the NPEs). Colleges and universities are now ramping up technology transfer offices to pursue patents in hopes of supplementing their endowments with licensing fees. In the meantime, academically based inventors are more motivated than ever to generate inventions for processing by the technology transfer offices because the generous provisions of the Bayh–Dole Act require that the inventors get a cut of the royalties.[3] Meanwhile, many large corporations go through tax law gymnastics to shield their IP-generated income from US taxes through the use of offshore entities.[4]

Of course, patent law firms remain at your disposal to serve as native guides to navigate the USPTO's byzantine array of patent laws, rules, and regulations, helping companies to obtain patents, providing legal opinions, and assisting with patent litigation. For those unable to afford the significant costs of patent litigation, there are now investment companies who will fund patent litigation in exchange for a portion of the booty. Technology businesses can also purchase IP insurance that provides a source of funding for IP litigation should the need arise. You can also hire a consulting company that is not a law firm to run and manage your patent litigation and/or manage the licensing activities for a piece of the action.

Welcome to the new world of IP, where options abound and the free market concept of *caveat emptor* is alive and well.

1.7 THE PERVASIVE NATURE OF IP INTERCONNECTIONS AND INTERACTIONS

The typical technology business has a number of basic interconnected functions. Figure 1.1 illustrates the interconnections among several key functions/departments of a typical technology business—namely, marketing, finance, human resources, product development, and manufacturing.

While the interconnections of the basic and more well-known business functions are generally understood, the interconnections between IP and the other business functions can go unappreciated. And yet IP has a direct role in each of these functions. In marketing, IP is used to protect product names using trademarks, to protect written material and designs using copyrights, and to boast about patents owned or pending. In finance, IP activities need to be closely budgeted because the amount of money available for IP dictates the company's IP strategy and the IP goals it can realistically pursue. In human resources, IP is generated

Figure 1.1 The IP interconnections between some of the main functions of a technology business.

by employees who perform in accordance with invention assignment agreements (at least in the United States) and who must balance the time they devote to IP with the demands of other activities and responsibilities.

In product development, IP is used to protect the inventions and designs embodied in products. In manufacturing, the inventions embodied in the products may be protected by one or more patents. Product clearance should generally be performed to ensure the product does not infringe the patents of others. Additionally, the products may benefit from trademark protection. While IP often happens behind the scenes and its day-to-day impact on individuals can be evanescent, its role in a technology business is ubiquitous.

In the course of conducting business, every technology enterprise needs to interact in one way or another with various external entities. Figure 1.2 illustrates some key IP-related interactions associated with running a technology business.

The IP interaction between a technology business and its suppliers often requires providing the suppliers with confidential information concerning present and future products. Nondisclosure agreements (NDAs) typically are used to ensure generally that such confidential information will remain protected—and, specifically, from a patentability viewpoint—that it remains publicly undisclosed. The IP interaction between a technology business and its customers may also involve the sharing of confidential information in both directions, and a mutual NDA is typically used to cover that two-way exchange. Customers may also improve upon a business's products and seek to patent the improvements, which can place the technology business in the awkward position of competing and negotiating with

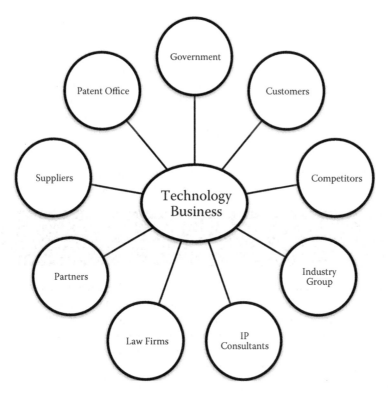

Figure 1.2 The IP interactions between a technology company and some example external entities.

its customers for IP ownership. The IP interaction between a technology business and its competitors can involve IP infringement on either side. Infringement can lead to litigation, licensing or cross-licensing agreements, or in some cases a standoff where nothing happens because neither side wants to rock the boat.

The IP interaction between a technology business and the USPTO is essential if the business is to obtain and maintain patents and trademark registrations. A technology business's IP interaction with the US government involves business regulations and can include IP-related antitrust issues. The IP interaction between a technology business and its law firm involves the seeking and obtaining of advice and counsel on such IP matters as obtaining patents from the USPTO, securing licensing deals, obtaining infringement/noninfringement and patent validity/invalidity opinions, navigating litigation, and so on. A technology business's IP interaction with consultants may involve getting business and other nonlegal IP-related advice and information on such matters as IP process improvement, prior art searches, and analytical evaluations of an IP space in the form of patent maps.

The IP interaction involving industry groups may include, for example, setting standards for the industry, forming patent pools for licensing IP related to industry standards, and seeking ways to formulate IP policies that benefit the industry as a whole. The IP interactions related to business partners can involve joint development agreements, dealing with issues of IP ownership for a jointly developed IP, how to share and leverage jointly owned IP, and so on. In some cases, so-called partners may also be customers. The partnership IP interaction is becoming both increasingly common and increasingly complicated as technology grows more complex and technology companies are forced to work more closely than ever to gain access to valuable and mutually beneficial IP and know-how. The partnership interaction almost always involves discussions and negotiations about "who owns the IP."

The processes and procedures that a technology business uses to handle its IP interconnections and interactions is what *IP management* is all about. A technology business's *patenting system* is a subsystem of its IP management system.

1.8 THERE IS NO ONE-SIZE-FITS-ALL PATENTING SYSTEM

While technology businesses may look similar from afar, each has its own organizational culture, values, and mix of personalities that make it unique. Different technology businesses operate with different strategies and at different levels of efficiency. Morale can vary from the high spirited and jovial to the outright depressed. Some have a high level of professionalism and are model corporate citizens, while others are ethically challenged and see lots of gray where most businesses see black and white. Some have strong executive leadership, while others are all but adrift. Some companies are excellent at implementing best-practice business processes and procedures, while others make up everything as they go along. Anyone who has worked at or closely with a number of technology companies can tell you that the differences between corporate personas can be truly startling.

A technology business should implement and operate its patenting system in a manner that plays to its particular strengths and avoids its inevitable weaknesses. The patenting system at IBM, for example, works well for IBM. It was developed by IBM over the course of 100 years of doing business, and it fits IBM's corporate culture and workforce. It was shaped by IBM's long history of innovation in computer technology as well as by some adverse history involving antitrust regulations.

So, while there may be some aspects of another company's patenting system that are enviable and worth emulating, it would be a mistake for a technology business with a different corporate culture, history, workforce, products, and budget to attempt to copy another's system too closely. Implementing an effective patenting system is a bit like personalized medicine in that it needs to be based closely on the corporate DNA to have a targeted effect.

NOTES

1. Oppenheimer and Snyder's other simplifying assumptions included that stars have a uniform density and pressure, no shock waves, no ejected matter, no radiation, and no rotation, none of which is true of a real star.

2. Invention promoters historically have marketed to independent inventors and not to technology companies. The fact that the USPTO has a standard complaint form for invention promoters speaks to their reputation.
3. See, for example, http://www.autm.net/Bayh_Dole_Act1.htm for information about the Bayh–Dole Act and its role in academic patenting.
4. See, for example, Dixon, Kim, "Microsoft, Hewlett-Packard shield billions from taxes, senator says," Reuters, last modified January 6, 2013, http://bottomline.nbcnews.com/_news/2012/09/20/13989718-microsoft-hewlett-packard-shield-billions-from-taxes-senator-says?lite

REFERENCES

Smith, A. 2000. *The Wealth of Nations*. New York: The Modern Library.
Thorne, K. 1995. *Black Holes & Time Warps: Einstein's Outrageous Legacy*, New York: W.W. Norton.

CHAPTER 2

THE IP UNIVERSE

2.1 IT'S MORE COMPLICATED THAN YOU THINK

One of this book's themes is that patenting in the IP Era of the Information Age is more complicated than most people working in the technology world realize. The IP universe is complex, and assumptions based on a spherical patent in a vacuum quickly break down and yield the wrong answers in important situations.

This leads us to an existential question: Why is the IP universe so complex?

2.2 THE IP BIG BANG

To answer this question, consider the cosmology of the IP universe. We can trace the origins of the IP universe in the United States to the IP Big Bang that happened on September 13, 1788, when the Continental Congress voted to put the US Constitution into effect after its ratification by the states. The US Constitution contains Article I, clause 8, section 8, which reads in part as follows:

> *Congress shall have [the] power...to promote the Progress of Science and useful Arts, by securing for limited Times to Authors and Inventors the exclusive Right to their respective Writings and Discoveries.*

This simple sentence gave birth to IP space in the United States. It enabled Congress to pass laws about IP and to set up a governmental patenting system operated by the United States Patent and Trademark Office (USPTO) as part of the Department of Commerce. The first patent statute Congress enacted was the Patent Act of 1790, which took effect on April 10 of that year. The first patent to enter IP space issued shortly thereafter on July 31, 1790, for an invention related to making potash. The IP space soon started filling up with patents while the laws governing the IP universe were improved upon by a number of subsequent patent acts passed by Congress (Chisum and Jacobs 1992).

Today the IP universe in the United States is filled with over 8.5 million patents and counting. It is governed by hundreds of sections of patent statutory law as embodied in Title 35 of the United States Code (USC). These statutory laws are sweeping and define generally what patents are and what rules govern their existence. In addition to these laws, there are hundreds of patent regulations embodied in Title 37 of the Code of Federal Regulations (CFR). These regulations seek to fill in the procedural details of how to operate within the statutory patent laws.

At an even finer level of detail are thousands of sections of patenting procedures as embodied in the *Manual of Patent Examining Procedure* (MPEP), which patent examiners at the USPTO use to examine patent applications. At last measure, the MPEP occupied two telephone-book-sized volumes with a combined thickness of four inches and a weight of about ten pounds. Rest assured that the updated editions will be thicker and heavier.

Residing within the interstices of this complex latticework of statutory laws, regulations, and procedures about patents is an immense body of patent case law. This case law covers thousands of court cases that apply the patent laws to particular sets of facts pertaining to patent-related disputes. The variations in the court decisions over time represent the ebb and flow of the legal tide, which can slightly shift or completely wash away existing law and in turn slightly or greatly impact the overall strength and enforceability of patents.

The passage of the Leahy–Smith America Invents Act in September 2011 (referred to in the remainder of the book simply as the "AIA") made significant changes to some of the key provisions of the patent laws. For example, it changed the United States from a "first to invent" country to a "first to file" (or, more accurately, a "first inventor to file") country, introduced new postgrant procedures, defined a new "microentity" applicant, and expanded the prior-use defense of a trade secret from just business-related inventions to all inventions.[1] Those of us who work in the patent field are waiting to see how effective some of the new provisions, such as the new postgrant review process, will actually be.

With respect to case law, of particular interest are patent-related court cases decided by the US Court of Appeals for the Federal Circuit (CAFC) and by the US Supreme Court. Patent law is federal law so patent court cases are first heard in federal courts—namely, the various district courts of the federal court system. The CAFC is the court where patent cases tried in the various federal district courts are appealed. The next and last step beyond the CAFC is the Supreme Court, which has the last word on what the law is and how it should be applied. The Supreme Court appears to be more willing than ever to take on cases related to patents, which as noted before, is a sure sign of the prominent role that patents are playing in our increasingly knowledge-based economy.

All of these laws, regulations, procedures, and case law decisions serve to warp the space–time continuum of the IP universe in a manner that would make Einstein proud. Classical Newtonian thinking doesn't always work, and in fact it more often leads to problems. All kinds of forces arising from the nonlinear interactions among and between the patent laws, regulations, and procedures as well as the nuances of the case law decisions operate on the intellectual matter that resides within IP space.

This makes the IP universe a very complex and fascinating place. Some parts of IP space are occupied by IP black holes where patents are so densely packed that not even a single photon of innovation can get close without being sucked in. Other parts of IP space have neutral zones where different high-tech businesses hold patents but don't enforce them for fear of retaliation. Technology businesses are expanding the frontiers of occupied parts of IP space by exploiting the *fractal nature of innovation*, a concept introduced and discussed in Chapter 8. The USPTO is constantly issuing patents that grant ownership to regions in IP space ranging in size from the atomic to the galactic, in areas of intense commercial interest in IP space as well as in areas where no one will ever visit for the lifetime of the IP universe.

Other forces can affect the shape of IP space. Some patents go supernova when the judicial system applies pressure. IP space wars are constantly being fought over regions of IP space where serious money is at stake. Consider the IP space for mobile phones, where there are so many lawsuits involving so many companies that a chart connecting the different companies by their lawsuits looks like a Manhattan subway map (just search the Internet for "smartphone lawsuit diagram" to check out half a dozen or so lawsuit diagrams). There are IP wormholes that allow an "obvious" invention to travel to another part of IP space where it becomes "nonobvious." All patents ultimately expire and turn into the cold, dark, and unenforceable matter that contributes to the publicly available regions of IP space called ***prior art.***

It should come as no surprise, then, that wandering into IP space and making all kinds of simplifying assumptions about how patents are supposed to work can cause problems, increases patenting system dysfunction, and often leads to disaster. There is no making up or guessing at the rules; over 200 years of lawmaking, rule-making, process refinements, and court cases already have established them. There are also well-understood IP best practices that savvy technology businesses know about and use every day to maintain an efficient patenting system.

People already have explored IP space and generally know what is out there and how it all works. It is a mistake not to learn from them.

2.3 BUSINESS ENTROPY: THE DISORGANIZATION OF THE ORGANIZATION

For any system to run efficiently—be it a government, a diesel engine, or a patenting system—it must adhere to the set of rules or processes that optimizes the system's operation. A system's level of efficiency is a function of the degree to which that system operates in an orderly and organized manner. One measure of a system's degree of order or disorder is entropy. And in the IP universe, entropy rules.

So it is worth taking a moment to consider what entropy is. The concept of entropy is rooted in the notion that of all the possible ways in which things may be ordered, there is a preferred or optimal order. The bedroom of a typical teenager is a perfect system for illustrating the concept of entropy. As much as it might strain the imagination, let's assume that the preferred and optimal order of things is when everything is put away in a designated place. Socks are in the sock drawer. Pants are in the pants drawer. The dresser drawers are closed. Shirts are hung up in the closet and ordered by color. Shoes are in the shoe rack and ordered by function (dress, casual, grunge, etc.). The bed is made. The shade on the bedside lamp is not crooked. The carpet is vacuumed. The top of the dresser is clear. The bedroom can be said to be in a state of low entropy.[2]

Now imagine the same bedroom after just 1 week of being subjected to habitation by said teenager without parental intrusion. No energy has been expended in keeping it clean and organized. Clothes are strewn everywhere. The bed is unmade. The lampshade is crooked. The drawers to the dresser are all ajar. The carpet is a mess. The shoes are in a random pile at the bottom of the closet. The bedroom has now descended into its more familiar state of high entropy. It is important to appreciate here the seemingly infinite number of ways in which the things in the bedroom might be in a place other than the designated places. The number of possible ways the bedroom can appear disordered far exceeds the relatively few ways in which the bedroom can be deemed "ordered."

Clearly it is much easier for the bedroom to exist in a state of high entropy than in a state of low entropy. It is as though the bedroom automatically drifts toward the high-entropy state. And in fact that is exactly what it does. The second law of thermodynamics states that systems tend to evolve toward increasing entropy. Not only that, but in physics entropy is expressed as a logarithmic function, which, as you will recall from high-school math, is a way of expressing large numbers by their exponents. The fact that entropy is usually expressed logarithmically tells us something about how large entropy can get in some systems, and thus why things can get so out of hand so fast. Consider, for example, that the number of possible ways to order a deck of cards is 8.066×10^{67}. You can see that trying to maintain the cards in one particular order could be an inexorable task.

Entropy applies as surely to business systems as it does to physical systems like bedrooms, collections of atoms, chemical reactions, and a deck of cards. A technology business's patenting system needs to have low entropy—that is, a high degree of order—for it to be effective and manageable. Achieving and then maintaining low entropy takes energy, which for a patenting system includes putting in place—*and then following*—well understood, but apparently not widely applied IP best practices. There is a right way—an *organized* way—of doing things as part of a patenting system just like there is a right way of putting things away in a bedroom and of making everything from semiconductor chips to potato chips. And because of the complexity of the IP universe, there are a huge number of ways to do things wrong. It is incredibly easy to achieve a patenting system with huge entropy quickly while spending a lot of money.

The more complex the system is, the more easily it can slide into high entropy. Because the IP universe is complex, controlling IP entropy is an insidious and daunting problem for technology businesses that are trying to run an effective patenting system.

2.4 TO HAVE ENTROPY OR NOT TO HAVE ENTROPY

Noncompliance with established procedures inevitably leads to system disorganization and a slide toward high entropy. The results are familiar—a messy room where you can't find what you need when you need it, nonfunctioning semiconductor chips that make people swear at their computers, and burned potato chips. Patenting systems with low entropy stay that way only when there is little tolerance for noncompliance with established best-practice IP procedures. It takes much more energy and work than most people think to set up and maintain a low-entropy patenting system.

That said, some technology businesses can and do get by with a simple patenting system that operates with relatively high entropy and yet still meets its patenting needs. It may be that a technology business needs patents only as a form of window dressing[3] or to hit a certain target number of patents (say, to impress investors). Or, it may be that the business needs only to produce patents covering inventions it doesn't even make or sell, with the patents having just enough strength to extract nominal licensing fees from someone who actually does make and sell products but would rather not get involved in patent litigation. Whatever the case, a patenting system with high entropy will almost always ensure that patent quality is wanting. But sometimes patent quality is not the highest priority of the patenting system. As we discuss later on, just like with tangible products, sometimes cheap and lousy patents can be good enough for the job.

At the other end of the spectrum are technology businesses that need patents of the highest quality, so they must have a patenting system with minimum entropy. A technology business whose patent strategy involves making and selling products and actively enforcing patents on those products via litigation needs patents that can withstand a rigorous legal challenge.

Many businesses assume that the USPTO patent examination process will weed out low-quality patents. As we will see below, the patent examination process is not as rigorous as most people assume it is. As a consequence, it is easy for a technology business to remain blissfully unaware of the fact that the patents its high-entropy patenting system is generating are shoddy to the point of being useless.

Because patenting systems are complex, they are subject to the functional indeterminacy theorem of complex systems, which states:

> In complex systems, malfunction and even total nonfunction may not be detectable for long periods of time, if ever (Gall 1975).

Most complaints about patents seem to come from those who yearn for the patenting game to be simple. They want it to resemble checkers, which perhaps it did in the early years of the IP universe. But here in the United States, the IP universe has been expanding at an accelerating rate for over two centuries. The technology economy and patent laws, rules, regulations, and court cases have all evolved to the point where the patenting game now resembles chess. It will never be like checkers again unless Congress finds a way to undo the IP Big Bang and creates a new IP universe from scratch. In the meantime, the IP universe as it now exists is complicated and unfair, just like the real universe.

The good news is that technology workers need not have a PhD in IP astrophysics to participate effectively in their organization's patenting system. However, technology workers do need to have an appreciation for the complexity of the IP universe, if for no other reason than to let go of preconceived and oversimplified assumptions about how it all works. The axiomatic goal of a patenting system is to generate patents with the assumption that the patents somehow provide a greater return than their expense. Implementing and operating a patenting system is for most technology businesses primarily an exercise in keeping entropy under control so that the patenting system can generate the kind of patents that provide the technology business with a good return on its investment.

2.5 MANIFESTATIONS OF PATENTING SYSTEM ENTROPY

Patenting system entropy manifests as a host of observable characteristics and behaviors. These characteristics and behaviors are rooted in the failure to follow IP best practices. Here are the top twenty-five manifestations to be on the lookout for:

1. Misalignment of patent claims relative to actual business needs
2. Patents of dubious validity when strong validity is intended
3. Lack of articulated central organizing principles and corresponding patenting strategies[4]
4. Unclear accountability and responsibility for patent-related matters
5. Patents with technically inaccurate descriptions
6. Patents that lack enabling details
7. Half-hearted provisional application filings

8. Scheduling crises and time crunches for patent-related matters
9. Low inventor morale
10. Paucity of cited prior art for patents in a known crowded IP space
11. Under-reporting of inventions and innovations to the business
12. Patents that have incorrect or uncertain inventorship
13. Business decisions about inventions being made by the inventors
14. Legal decisions about inventions being made by inventors and/or businesspeople
15. Business decisions about inventions being made by lawyers
16. Technical decisions about inventions being made without inventor input
17. Half-hearted review of patent applications by inventors
18. Limited technology-worker participation in the patenting process
19. Abandoning patent applications that should have never been filed
20. Legal bills that are disproportionate with respect to patent value obtained
21. Long delays from innovation generation to patent application filings
22. Inventorship politicization
23. Poor or nonexistent communication channels
24. Lack of documented procedures that govern the patenting system
25. High volume of patent-related misinformation and misguided assumptions

It is worth discussing in some detail the root causes of patenting system entropy so that they can be addressed and remedied at a fundamental level.

2.6 HOW HARD CAN IT BE?

The failure to follow or appreciate the key steps in patenting system processes is often the result of a "how hard can it be?" mind-set. This mind-set assumes a spherical patent in a vacuum. As a result, it makes too many oversimplifying assumptions about how to integrate a patenting system effectively with other functions of the business as well as how to manage the full dynamic range of operational details. The usual result of this mind-set is a spike in patenting system entropy because the disorder propagates through the patenting system and generally leads to patenting mayhem.

The "how hard can it be?" mind-set reflects a lack of appreciation for the second law of thermodynamics and just how fast entropy can drive a system to a state of utter dysfunction. It also betrays a lack of awareness about just how many more nonfunctioning states a system can have than functioning states that represent the right way to do something.

In patenting, the "how hard can it be?" mind-set often comes with watching how other technology businesses get patents and wanting to imitate them. A gallery of framed patents on the lobby walls of a technology business can look really impressive. Technology businesses that are successful at patenting make it look easy. However, unless one has worked as part of a well-honed, process-oriented, and disciplined patenting system, it can be difficult to appreciate the enormous behind-the-scenes effort, planning, organization, procedures, and cost that it takes to get the details right and maintain order. For executives that tend to think in big-picture terms and let others sweat the details, the "how hard can it be?" mind-set is an especially seductive trap.

2.7 LOGIC IS DANGEROUS

The lack of real IP education and training for technology workers leads to another more insidious problem: technology workers having to fill the knowledge vacuum with assumptions they make about IP matters based on logic and common sense. Unfortunately, the nature and character of IP and of patents, in particular, are defined by the underlying patent laws, which are nonintuitive and do not lend themselves to logical deduction. Justice Oliver Wendell Holmes said it best:

> *The life of the law has not been logic; it has been experience.... The law embodies the story of a nation's development through many centuries, and it cannot be dealt with as if it contained only the axioms and corollaries of a book of mathematics. (Holmes 1881)*

If Holmes leaves you feeling skeptical, let's shift the context. Imagine doing your income taxes based on logic rather than on the rules (that is to say, the laws) of income taxation as set forth in Title 26 of the United States Code. What penalty do you think you would owe Uncle Sam after taking a logic-based approach to income tax preparation?

This resistance of patents to the application of logic and common sense can be difficult to accept. As a result, there is a staggering amount of IP misinformation in the high-tech workplace. This misinformation can be quantitatively measured in the number of IP BITTs that appear in IP-related conversations and e-mails. A "BITT" is a sentence that starts with the phrase "But I thought that…" Here are the top twenty IP BITTS I hear all the time:

1. BITT for the invention to be patentable you had to have a working model.
2. BITT we could keep that part of the claimed invention a secret.
3. BITT she was an inventor because she did experiments that verified our results.
4. BITT we couldn't infringe that patent because we were using that invention before they were.
5. BITT we couldn't infringe that patent because we invented it first.
6. BITT we couldn't patent that!
7. BITT that patent was invalid because it's obvious.
8. BITT for it to be patentable the invention needed to work well enough to be commercialized.
9. BITT we couldn't infringe that patent because our device uses more elements than are claimed in the patent.
10. BITT we didn't need to include details about how the invention works because it would limit our claims.
11. BITT we couldn't be an infringer because we don't make all the components of the final device that the patent claims.

12. BITT writing up a description of the invention would take a week and I don't have that kind of time.
13. BITT I didn't really need to read the patent application for my invention because it's all legalese anyway, so I just signed the filing documents.
14. BITT we were supposed to search for prior art relating to the invention.
15. BITT we wouldn't infringe because we don't perform the method steps in the order presented in the claims.
16. BITT we could just add all of that new information after the application was filed.
17. BITT we would be better off not citing that reference to prior art because it would prevent us from getting the patent!
18. BITT a provisional patent [*sic*] doesn't need all of those details.
19. BITT a patent cooperation treaty (PCT) patent [*sic*] would give us world-wide patent protection.
20. BITT there was no prior art to my invention because I searched the Internet and I didn't find anything.

There are giga-BITTs' worth of this kind of patent misinformation, all of which is easily generated through the application of logic and common sense. Much of this misinformation is based on the twin assumptions of a spherical patent and a simple IP universe.

NOTES

1. For an overview of the AIA, see, for example, Gortych, Joseph E. 2012. US patent reform and photonics companies. *Optics & Photonics News,* June 2012: 16–17; for the gory details, see Brinckerhoff, C. et al. 2012. *America Invents Act—Law and analysis,* rev. ed. Alphen aan den Rijn, Netherlands: Wolters Kluwer Law and Business.
2. Which of course must mean that the teenager has been away and someone has gone in and cleaned the room.
3. "[P]eople will obtain a patent simply for the glory of hanging a beribboned document on the wall." Pauline J. Newman, concurring in *Figueroa v. United States,* 05-5144 (Fed. Cir. 2006).
4. Chapter 13 discusses these topics in more detail.

REFERENCES

Chisum, D.S. and Jacobs, M.A. 1992. *Understanding Intellectual Property Law.* New York: Matthew Bender.
Gall, John. 1975. *Systemantics: How Systems Work and Especially How They Fail.* New York: Quadrangle/New York Times Book Co.
Holmes, O.W. 1881. *The Common Law.* Boston: Little Brown and Company.

CHAPTER 3

BEYOND THE SPHERICAL PATENT

3.1 SURFACE FEATURES

All patents have in common a number of surface features: a title, an abstract, a field of the invention section, a background of the invention section, a summary of the invention section, a brief description of the drawings section, a detailed description of the invention section, and a claims section. Almost all patents include black-and-white line drawings depicting various aspects of the invention.

Because of their common layout, all patents tend to look the same whether they are incredibly valuable or absolutely worthless. The scrawniest patent with just a few pages and a couple of mundane figures could be an absolute bonanza, while the 100-page behemoth with dozens of intricate apparatus figures, fancy flow diagrams, impressive equations, and a hundred claims can be worth less than the paper on which it is printed. All patents are spherical when you look at their surface features only (Figure 3.1).

You can't judge a patent by its cover page. In fact, you can't even judge it by reading it—not at least in the normal (that is to say, nonlegal) way one reads a newspaper or a technical journal. This is mainly because patents are written in an obtuse style and language informally called *patentese*. Patentese is a weird hybrid of technical and legal jargon whose purpose is to provide an exacting technical and legal description of an invention. However, like a foreign language, it can be rather confusing and mysterious, especially when it comes to trying to understand detailed concepts like those expressed in claims. Just as French can be used make the commonest of foods sound gourmet,[1] patentese can make the most trivial invention sound profound.

Here is one version of what a pencil sounds like in patentese:

> *An instrument for marking a surface with graphite, the instrument comprising a cylindrical member having a polygonal cross-section, a frustoconical proximal end, an opposite distal end, and an axial bore. A cylindrical graphite member is fixedly supported in the axial bore and has a substantially pointed end that extends from the frustoconical proximal end of the cylindrical member. The instrument includes a deformable element secured at the distal end of the cylindrical member. The deformable element comprises a material that substantially removes said graphite marking from said surface when the deformable element is frictionally and deformably applied to said graphite marking.*

In many ways, patents are like diamonds. The fancy official-looking patent document written in patentese can appear dazzling to the ingénue. However, truly understanding a patent's quality and substance requires a jeweler's eye.

Figure 3.1 A spherical patent and its surface features.

3.2 WHAT LIES BENEATH

Figure 3.2 on the next page shows two spherical patents being smashed together to reveal eleven main requirements hidden beneath a patent's surface. These requirements are what really matter when evaluating a patent or patent application for a given invention.

3.2.1 Statutory subject matter

The *statutory subject matter* requirement maintains that only certain types of inventions are patentable, as summarized in Table 3.1. This table assumes a spherical patent and gives a first-order view of what is patentable and what is not. As you can imagine, there are subtleties associated with each listing. For example, while compositions of matter already existing in nature are not patentable, methods of extracting or purifying such compositions can meet the statutory subject matter requirement. Also, certain articles of manufacture that are purely ornamental are the subject of "design patents," as opposed to the more well-known "utility patent."

Moreover, an invention must produce a useful, concrete, and tangible result in order to qualify as statutory subject matter. A process is likely to be deemed statutory subject matter if it is tied to a particular machine or apparatus or if it transforms a particular article into something else. Business methods as well as most other types of methods can be considered statutory subject matter if they qualify as a true process rather than just an abstract idea. Inventions that cover an entire "human organism" do not qualify as statutory subject matter. Naturally occurring DNA sequences are "products of nature" and thus do not qualify as statutory subject matter.

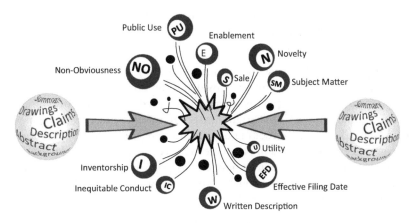

Figure 3.2 Two spherical patents being smashed together to reveal their underlying main requirements.

Table 3.1 Statutory subject matter

Patentable	Not patentable
• Processes/methods	• Laws of nature
• Systems/apparatus	• Abstract ideas
• Articles of manufacture	• Compositions of matter that exist in nature
• Compositions of matter	• Pure mathematical relationships (e.g., $E = Mc^2$)
	• Inventions deemed against public policy

3.2.2 Novelty

The *novelty* requirement mandates that the invention be something new. This requirement is measured with respect to the invention as it is *claimed* in the patent, not with respect to the invention as it is *described* in the patent. Novelty is established at the effective filing date of the patent application and not at the time the invention occurred. This reflects the recent shift in the US patenting system from a first-to-invent to a first-to-file system.

The AIA deems that inventions directed to tax strategies by definition fall within the prior art—meaning that by definition they fail the novelty requirement.

One way to establish an invention's lack of novelty is by finding a description of the invention in a single prior art reference covering, well, the same invention. This lack of novelty is also called "anticipation," and the single prior art reference is said to "anticipate" the claimed invention.

3.2.3 Effective filing date

Every patent application and patent has an *effective filing date,* which is the earlier of (a) the actual filing date of the patent or patent application or (b) the filing date of a patent application from which the earliest priority can be properly claimed (i.e., it must support the claims that are being asserted in the later application). That earlier patent application can be a US provisional application, a US nonprovisional application, or a non-US patent application. The effective filing date is important because it is the date that (with some exceptions) generally describes what constitutes *prior art*, which is any available information related to and that predates the invention.

Assuming a spherical patent, any information relating to the invention that arises *subsequent* to the effective filing date by definition does not constitute prior art, while information relating to the invention that predates the effective filing date does constitute prior art. This bright line is a result of the recently implemented first-to-file rule. However, looking beyond the spherical patent approximation, there are exceptions. For example, there are 1-year grace periods during which an inventor can disclose the invention (e.g., by publishing information about the invention), use the invention publicly, or offer the invention for sale prior to filing a patent application on the invention.

The effective filing date is not a requirement per se. Every patent application and patent has an effective filing date. However, it is considered a requirement for our purposes because it needs to be established properly in order to enjoy its potential benefits. As discussed later in Chapter 6, there is tremendous confusion about how to establish an effective filing date properly using a provisional patent application. Under the first-to-file system, when a person *other than the inventor* files a patent application, that patent application has a bogus effective filing date because a patent cannot be granted to someone who is not an inventor. This can happen when someone "derives" (borrows, steals, uses, etc.) the invention of another and then tries to patent it. The USPTO accounts for this possibility by allowing an inventor with a later filing date to present evidence in a "derivation proceeding" that an applicant with an earlier filing date derived the claimed invention from them, the "real" inventor.

There are many ways an effective filing date can be established. Here are a few example effective filing dates:

- The filing date of the patent application that later issues as a patent
- The filing date of an earlier filed non-US patent application that is followed within 1 year by the filing of a regular US patent application[3]
- The filing date of a provisional patent application in the United States that is followed within 1 year by the filing of a regular patent application
- The filing date of a so-called "parent application" from which a "continuation application" was filed that discloses the same subject matter but with a different set of claims

A patent can have more than one effective filing date. For example, a patent can claim priority to multiple provisional patent applications that were previously filed. So, then, which effective filing date is the one that establishes what constitutes prior art for the invention? The answer is that the effective filing date will depend on what is being claimed. Different claims can have different effective filing dates. In such cases, one needs to read the claims and then look at all of the related patent application filings. The patent document does not always tell you which claims are entitled to which effective filing date when there are multiple effective filing dates. You have to figure that out for yourself.[4]

The effective filing date has increasing importance and may be subject to increased scrutiny by a competitor if a US nonprovisional patent application claims priority from an earlier filed application (e.g., provisional or international application) and relevant prior art shows up in between the effective filing date and the filing date of the nonprovisional patent application. We discuss this scenario at some length in Chapter 6 in connection with provisional patent applications.

With the advent in the United States of the first-to-file system, many inventors, in their rush to be the first to file, may fall into the trap of filing a patent application that does not adequately disclose the invention as they would like to or need to claim it later on. Are they still entitled to the effective filing date just because they were the first to file? What if the patent application is written on a paper towel in Pig Latin? Clearly there are limits to the paucity of disclosure that establishes a legitimate effective filing date.

3.2.4 Nonobviousness

An invention cannot be an obvious variant of a known (previous) invention; in other words, it must be **nonobvious.** Again, it is the invention as it is *claimed* that must meet this requirement. The nonobviousness requirement is very important and can be a source of confusion as well as contention. It is complicated and a source of angst for many involved in patenting.

The nonobviousness requirement is worthy of an extended discussion and that discussion is relegated to the entirety of Chapter 10.

3.2.5 Utility

The **utility** requirement is perhaps the hardest requirement to not meet. Here, utility requires only that the invention be operable for some beneficial purpose and that it not be immoral or against public policy.[2] This requirement comes in handy for the patent office when examining the patentability of ridiculous inventions such as perpetual motion machines, which are known to be impossible to build (and thus inoperable) by virtue of the laws of thermodynamics.

The utility requirement can prove problematic for certain pharmaceutical compounds that cannot be shown to have some practical effect or for chemical compounds that have no apparent use. Mostly, the utility requirement serves as a bad-invention rejection filter with a minimal threshold. If an invention operates in the way described and claimed in the patent application and can be shown to be nominally beneficial to society in some way, then it will in all likelihood satisfy the utility requirement.

3.2.6 Enablement

The **enablement** requirement demands that the description of the invention be set forth in sufficient detail for someone skilled in the subject matter of the invention (i.e., a "person skilled in the art") to practice the invention without undue experimentation. This requirement is why a patent needs to include actual details rather than ambiguous phrases and empty statements.

Enablement is directed to teaching someone else how to make the claimed invention. Enablement problems arise when the detailed description section discusses the invention's many virtues, *what* the invention can do, and *why* it does it, but never really gets around to teaching *how to make and use the invention.* A patent provides the owner with a time-limited monopoly on the invention with the trade-off that the patent disclose how to make and use the invention—that is, an enabling disclosure.

An interesting aspect of enablement is that you need not know the underlying phenomenon that makes the invention work. Rather, the description needs only to provide enough details for someone else to make and use the invention. For example, if someone discovers that sprinkling flour on a computer chip makes it run faster, there is no need to teach about the flour semiconductor physics at work. It is only necessary to teach how to add the flour, in what amounts, and so on, so that someone else can achieve the same result. The teaching can be the result of empirical evidence based on experiments with different types of flour, different amounts, annealing temperatures (perhaps "baking temperatures" is the more appropriate phrase here), and so on. That said, it is usually the case that when the underlying physics of the phenomenon is understood, the invention can be described and claimed in a more general way.

It is important to appreciate the difference between an idea and an invention. An invention is *enabled* while an idea need not be. For example, the transporter device used on the television series *Star Trek* to beam people and things from one place to another is just an idea and not an invention because you cannot actually teach someone else how to make and use it. Note that you could talk at length about what the transporter does, why it does what it does, and all of its wonderful advantages without ever getting into how it actually works or how to make it.

We can assess a given invention's level of enablement by playing the *How Game*, which involves repeatedly inquiring about *how* to make the invention and seeing where the increasingly narrow inquiry leads. Let's play the *How Game* with the *Star Trek* transporter and see what happens:

Q1. *So, Scotty, how does the transporter actually work?*
A1. *Well, a person stands on a transporter pad, and when I slide these switches down to the "transport" setting, the person gets sent to another location. It helps us get to remote locations really fast.*
Q2. *But that's what the transporter does and why you use it, not how it works. So again, how does it work?*
A2. *Oh, I see. Well, the person's molecules are disassembled at the first location and then beamed to another location where they are reassembled in essentially the identical configuration, give or take a few molecules.*
Q3. *But how does the device actually go about disassembling and reassembling the person's molecules without actually killing the person?*
A3. *Well, we haven't really figured that part out yet.*
Q4. *And how do you create the beam that transports the disassembled molecules?*
A4. *Well, come to think of it, we don't know that either. C'mon, it's just a TV show…*

By directing just a few simple "how" questions to fundamental aspects of the invention, you rather quickly get to "I don't know" or an evasively mumbled equivalent. This line of questioning reveals that the transporter is not an enabled invention but purely an idea and as such is unpatentable.

Playing the *How Game* with inventors often sends them back to the drawing board to work out additional details for certain aspects of their invention.

When I worked as an engineer and was inventing things and filing innovation disclosures, I was on the receiving end of many *How Game* questions. They always led me to a much better understanding of the invention, which in turn made for a better application and subsequently a better patent.

The *How Game* is good to play with new inventors who are still coming to grips with how much information is needed to satisfy the enablement requirement. Sometimes the *How Game* can end up sounding something like this:

Q1. *Well, Bob, you said that the light beam in your invention is modulated to provide an optimum intensity distribution. How is it optimized?*

A1. *We use a unique optimization algorithm based on a feedback measurement. The algorithm is the key to the invention.*

Q2. *But you don't say in your description anything about how the algorithm actually works. How does the algorithm work, Bob?*

A2. *We can't disclose that; it's proprietary!*

This version of the *How Game* illustrates the common situation where satisfying the enablement requirement runs up against the need to keep things secret. Patents are about disclosing the invention to the public in exchange for a mini-monopoly on that invention. It is difficult if not impossible to meet the enablement requirement for a patent if you need to keep a key piece of the invention secret. It is best to identify these kinds of proprietary information problems before sending a patent attorney off to draft the patent application.

Here is one more very important aspect of enablement that tends to be lost even on some patent attorneys:

> **A claim must be enabled to its full scope.**

In an effort to obtain the broadest coverage, claims are sometimes drafted in a manner that omits key components that are required to make the invention work. As we will see in our discussion of the relationship between infringeability and validity, omitting claim limitations broadens the claim and increases the chances that someone will infringe that claim. However, the downside of broadening the claim by omitting key limitations is that enablement for the claim can become problematic. The recent trend in the Court of Appeals for the Federal Circuit (CAFC, the highest patent court below the Supreme Court) is not to read limitations into a broadly worded claim in order to make it valid. Rather, the claim is taken at face value. If infringement is found, the patent is then examined to ascertain its validity by assessing whether the problematic claim is enabled to its full scope.

3.2.6.1 Case study: enablement problem with an overly broad claim

The following example, which is based on an actual CAFC court case, illustrates how enablement issues arise when a claim is overly broad relative to the detailed description of the invention. The court case involved US Patent Nos. 5,456,669 (the '669 patent) and 5,658,261 (the '261 patent), which are directed to injectors

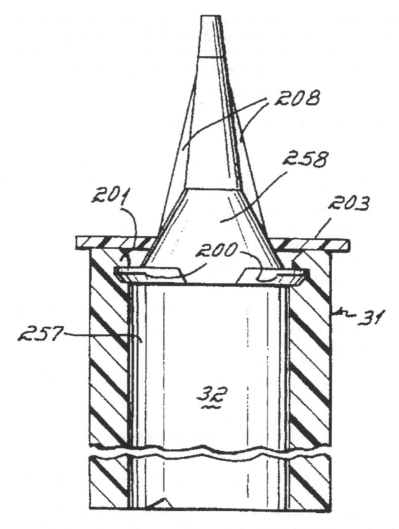

Figure 3.3 A front section of an injector device as depicted in Figure 15 of the '669 patent, showing a syringe 32 supported within a pressure jacket 31.

that use disposable front-loadable syringes.[5] Figure 3.3 is Figure 15 from the '669 patent. The specifications of the '669 and '261 patents are essentially identical.

Each of the embodiments of the injector described in the two patents includes a pressure jacket 31 into which the syringe 32 is inserted. The pressure jacket surrounds the syringe and prevents it from breaking under the internal pressure generated when a contrast agent within the syringe is injected into a patient.

The detailed descriptions of the two patents call for the syringe to have a pressure jacket. However, the patent owner found out that a competitor was using a similar syringe that did not include a pressure jacket. The patent owner then sought modified claims during prosecution of the patent application that omitted the pressure jacket limitation so that the competitor would infringe the claims. The examiner allowed these modified claims.

The case eventually reached the CAFC on appeal. The CAFC found that the competitor's syringe did in fact infringe the modified claims. However, the CAFC then found the modified claims to be invalid due to lack of enablement.[6] Specifically, the CAFC found that the modified claims were not enabled to their full scope because the patent did not adequately teach how to make a syringe that did not have the pressure jacket. The CAFC also noted that the two patents included undermining statements in their background of the invention sections that admitted that, without a pressure jacket, syringes that are able to withstand such high pressures are "expensive and therefore impractical where the syringes are to be disposable."

The owner of the two patents tried to argue that since the patents provided an enabling disclosure with respect to a single preferred embodiment (i.e., the syringe with the pressure jacket), the broader claim should be considered enabled. The CAFC rejected this argument and held that *the full scope of the claim must be enabled,* stating:

> *The irony of this situation is that [the patent owner] successfully pressed to have its claims include a jacketless system, but, having won that battle, it then had to show that such a claim was fully enabled, a challenge it could not meet. The motto "beware of what one asks for" might be applicable here.*

3.2.7 Written description

The **written description** requirement ensures that the inventors actually invented the invention set forth in the claims. Another way of thinking about the written description requirement is as proof that the inventors were in possession of the claimed invention at the time the patent application was filed (or, more accurately, by the effective filing date being asserted).

This may sound obvious, but situations can arise where the scope of the claims gets out of hand relative to the invention that is actually described. For example, consider a patent application for an optical system invention where the description, the drawings, and the claims consistently state that a thin-film **transmissive optical filter** is an element in the optical system and no alternative element is mentioned.

Later on, while the patent application is still pending, the patent owner finds out that a competitor's optical system is identical except that the competitor uses a **reflective diffraction grating** instead of a thin-film transmissive optical filter. The patent owner tries to amend the claims to include a reflective diffraction grating used as an optical transmission filter even though the reflective diffraction grating is not described anywhere in the inventor's patent application. The sole purpose of the change is to make the competitor's optical system infringe the inventor's claims if the patent issues. In such a case, the examiner could rightfully assert that the amended claim does not satisfy the written description requirement because a reflective diffraction grating is not described.

Note how this case differs from the case involving the pressure jacket and the syringe. In the former case, the patent owner tried to remove from a claim an item that the description said had to be there. In that case, the problem was lack of enablement. In the case of the reflective diffraction grating, the patent owner tried to add a limitation that was nowhere to be found in the patent application— hence a written description problem.

An interesting feature of the written description requirement is that one cannot argue that the missing description would be obvious in view of the existing description. In the case of the reflective diffraction grating, for example, one cannot argue that using it is an obvious variant of the transmissive grating system and therefore can be cited in the claims. As it was not part of the original written description, the examiner could rightly conclude that the inventors did not have possession of this aspect of the invention and thus were not entitled to claim it.

A more blatant example of a written description problem is when a claimed element is simply nowhere to be found in the description and is not even related to an existing element. This situation goes beyond our reflective diffraction grating example (wherein the owner tried to replace the described element with an equivalent), so it is easier to spot.

3.2.8 No public use

In the United States, an invention is not patentable if it has been in public use for more than 1 year anywhere in the world. Put differently, from the date of first public use, an inventor has up to 1 year to file a patent application, at which point the invention becomes unpatentable. Public use is based on the notion of the public disclosure of the invention. One can prevent the public disclosure of an invention by disclosing the invention under the aegis of a nondisclosure agreement. So the term *no public use* really means that there can be some public use, but that public use is circumscribed.

Public use does not necessarily mean that the invention is out in the open in the ordinary sense. The ordinary use of a machine invention in, for example, a store can constitute public use even though the internal workings of the machine are hidden. Public use is assessed based on whether the invention was ready for patenting at the time of the public use as well as the totality of the circumstances surrounding the public use. One such circumstance is whether the inventor had a reasonable expectation of privacy and whether he or she retained control of the invention.

Often, an assertion of public use is countered by the demonstration that the use was experimental in nature—namely, that the invention was not yet ready for patenting. Thus, in cases where an invention is used for experimental purposes, it is useful to document such use and to distinguish it from outright public use.

The new laws under the AIA appear to emphasize the public disclosure of the invention, and at this writing there is still some debate about whether public use requires actual disclosure of the invention. This is a perfect example of how the patent laws can shift and create a period of uncertainly in how the new law will be applied. For example, it was clear in the pre-AIA days that private (i.e., secret) commercial use of an invention started the clock ticking on a 1-year time limit to file a patent application on the invention. The AIA appears to have removed this commercial use limitation so that private exploitation of an invention for commercial gain (say, as a trade secret) does not preclude patenting the invention years later.

3.2.9 No sale

An invention is not patentable if it has been for sale or has been offered for sale for more than 1 year anywhere in the world in a manner that discloses the invention. Put differently, from the date of a sale or offer for sale, one has up to 1 year to file a patent application, at which point the invention becomes unpatentable.

The relevant part of the patent laws that was updated by the AIA is 35 USC §102(a)(1) states that (emphasis added):

> *A person shall be entitled to a patent unless—*
> *(1) the claimed invention was patented, described in a printed publication, or in public use, on sale, **or otherwise available to the public** before the effective filing date of the claimed invention;*

One interpretation of this is that the "or otherwise available to the public" means that the sale must publicly disclose the invention. Another interpretation is that this phrase does not pertain to the on-sale language upstream of this phrase and that the terms "public use" and "on sale" have already been construed under the existing law and courts will likely continue to construe them the same way under the AIA.

3.2.10 Inventorship

Patent applications must be filed in the names of the true inventors. You cannot add bogus inventors and you cannot leave out legitimate inventors. **Inventorship** is not the same as ***authorship.*** It is a legal requirement whose definition is much tighter than that of authorship. A spherical patent rule of thumb for what makes a person an inventor on a patent is that *a true inventor will have contributed to at least one claim.* Note that inventorship can change during the course of the patent examination since the scope of the claims can change (e.g., by amending them or canceling them).

Sometimes many people work on an invention even though, technically speaking, only one or two of them are inventors. This situation can result in pressure to list people as inventors just because they assisted with the development of the invention. Chapter 11 discusses inventors and inventorship in greater detail.

It is inappropriate to add as inventors people who are not true inventors or to omit actual inventors. It can lead a patent to be declared invalid and unenforceable. Nevertheless, office politics can often play a role in deciding who is and is not an inventor. There are quite a few technology businesses where managers seek to have themselves included as inventors on patent applications as a matter of routine regardless of whether they made any contribution to the invention.

3.2.11 No inequitable conduct

Inequitable conduct is a defense to patent infringement that can render the entire patent unenforceable. It has been called the "atom bomb" remedy of patent law because it blows up the entire patent and all the claims, not just certain claims as in a standard invalidity determination. The inequitable-conduct standard has its roots in the so-called unclean-hands doctrine that has been used in the past to dismiss patent lawsuits where egregious misconduct (e.g., outright fraud) was used to obtain the patent. So a requirement for a patent is that there can be *no inequitable conduct* with regard to how the patent was procured. As you might suspect, some thresholds need to be crossed before a particular conduct is considered "inequitable."

Until fairly recently, inequitable conduct was determined on two main bases: the presence of the intent to deceive and the materiality of that deception in allowing the patent to issue. This determination used to be based on a kind of sliding scale. As the degree of materiality increased, the level of intent needed to be demonstrated decreased. In some cases, negligence or gross negligence could meet the intent standard if the materiality of the deception was sufficiently high. The courts believed that by reducing the standards for intent and materiality (i.e., by not requiring that both be demonstrated), they could encourage applicants to be more forthright with the USPTO.

All those associated with the filing and prosecuting of a patent application have a duty to deal with the USPTO with candor and in good faith. This duty includes disclosing prior art to the USPTO that is material to the patentability of the invention. The duty of disclosure can be thought of as the "no hiding the ball" provision. There is no hiding the ball—not by the inventors, the attorney, or anyone else involved in preparing the patent application—with respect to known prior art that could impact the examiner's review of the patent application.

Likewise, there is no hiding the ball with respect to information about potential bars to patentability based on public use and sale activities of the invention. An inventor must not hide the ball by falsifying data or results or by intentionally providing information about the invention that is untrue. Signatures on the patent application assignment must not be forged. There is a long list of ways to hide the ball, and any one of them could result in a finding of inequitable conduct on the part of someone involved in the patenting process.

In patent litigation, invalidity of the patent based on inequitable conduct is frequently asserted because there is often a good chance that at least some shred of evidence can be found that "suggests" inequitable conduct. The temptation not to disclose information that might be adverse to one's interest is just too great for some to pass up, especially if the chances of the USPTO catching one in the act are slim. For some people, getting a patent now and worrying about the fallout of any inequitable conduct allegations later is better than not getting a patent at all. However, litigators in the position of defending their client against a charge of patent infringement will endeavor to sniff out any hint of impropriety and seek to amplify it into inequitable conduct to try to knock out a patent.

It is worth reiterating that when a court finds that a patent was issued through inequitable conduct, select claims of the patent are not found invalid but rather the *entire patent* is ruled *unenforceable.* This is an important distinction because finding one claim in a patent invalid does not invalidate the entire patent. The other claims may still be valid and they might be infringed. However, a finding of inequitable conduct, even if the conduct related to only a *single claim,* shoots down the patent. This is why the charge of inequitable conduct is so often levied against the patent owner. Even in a Hail Mary situation, the small chance of detonating the inequitable conduct atom bomb is worth a try. It's amazing what one can uncover in discovery and in depositions about all the events that transpired during the prosecution of a patent that is not apparent from the cold, dry record of the documents in the USPTO.

The law governing inequitable conduct provides yet another example of how the patent laws can shift with time. The sliding scale standard discussed earlier that balanced intent with the materiality of the infraction has since been replaced with a more stringent standard.[7] Today, to prevail on a claim of inequitable conduct, the accused infringer must prove by clear and convincing evidence that the patentee acted with the specific intent to deceive the USPTO. In addition, the accused infringer must also show the materiality of the transgression to meet a "but for" standard. That is to say, but for the alleged infraction, the USPTO would not have granted the patent. There is an exception to this standard. Egregious misconduct, such as filing an unmistakably false affidavit, is automatically considered material without regard to the "but for" standard.

Why the change in the standard? Because the CAFC grew weary of patent infringers always playing the inequitable conduct card. The consequences of an inequitable conduct finding are so drastic that in many instances they were more inequitable to the patent owner than to the infringer. Moreover, the CAFC noted that relaxed standards for establishing inequitable conduct were promoting the overdisclosure to the USPTO of marginally relevant references on the part of the patent owners (or, more likely, their patent counsel).

It is also worth noting that the AIA addresses the issue of inequitable conduct by allowing a patent owner to ask for a supplemental examination of a patent after it issues to correct possible inequitable conduct issues as well as other potential or actual defects in the patent. A benefit of the supplemental examination is that the AIA specifically states that a patent cannot be held unenforceable based upon information that was "considered, reconsidered, or corrected during a supplemental examination of the patent."

3.3 THE EXAMINER FIELD OF VIEW

When examining patent applications, patent examiners have a limited amount of information available to them. That is to say, an examiner examines a patent application based on information that is within his or her field of view (Figure 3.4).

On this basis, we can divide the eleven main requirements into one set that is within the examiner's field of view and one that is outside the examiner's field of view (Table 3.2).

Figure 3.4 The patent examiner's field of view with respect to the main requirements.

Table 3.2 The examiner field of view

Within the field of view	Outside the field of view
• Statutory subject matter	• No public use
• Novelty	• No sale
• Nonobviousness	• Inventorship
• Utility	• No inequitable conduct
• Enablement	• Effective filing date
• Written description	

For example, the statutory subject matter requirement is within the examiner's field of view because it speaks to the nature of the invention as it is described and claimed in the patent application. The examiner is thus well positioned to compare the invention to the statutory requirements laid out in the MPEP in section 2106. This is not to say that the examiner will always make the right call on some of the inventions that are on the margin, but he or she will at least be able to make a reasonable judgment based on the information at hand.

The novelty requirement is also generally within the examiner's field of view because some form of prior art is used to compare the claims to. The same is true with the nonobviousness requirement. The enablement requirement and the written description requirement are also generally within the examiner's field of view because these requirements refer to properties that are inherent in the patent application. The effective filing date requirement is also generally within the examiner's field of view because this information is in the record of the patent application and the other filing documents that come before the examiner. The utility requirement is also generally within the examiner's field of view because it speaks to the operability of the described and claimed invention as set forth in the patent application.

But what about the requirements that limit the public use and the sale of the invention? The examiner will generally not have any information about such uses unless it is provided by the applicant or is so available in the public domain that it falls into the examiner's lap. Otherwise, these kinds of activities are beyond what an examiner can usually glean from the written record.

The same is true with the no-inequitable-conduct requirement. How would a patent examiner ever know whether an inventor was intentionally hiding something important (i.e., material) to the examination of the invention? For the inventorship requirement, how would a patent examiner know who the actual inventors are without interviewing them and conducting a fact-specific inquiry? A patent examiner cannot realistically be expected to assess whether these particular requirements relating to public use, invention sale, inventorship, and inequitable conduct are satisfied.

Moreover, the requirements that generally fall within the examiner's field of view can often drift outside it. For example, if the examiner is reviewing the novelty or the nonobviousness of the invention and the most relevant prior art is not actually in front of the examiner, then the real prior art is outside the examiner's field of view. This means that the novelty and nonobviousness requirements are not going to be assessed properly. Further, if a provisional patent application is used to obtain an effective filing date that it cannot actually support, then the requirement relating to having a properly established effective filing date has drifted out of the examiner's field of view.

An examiner's ability to examine the requirements normally within his or her field of view can be illusory depending on the quality of the information the examiner has at his or her fingertips. In the case of a provisional application that does not support the sought-after effective filing date, the information is actually within the field of view but may be of too fine detail for the examiner to notice easily. The examiner might have to scrutinize the provisional application filing and compare it word for word to the regular application filing to make this assessment. This is rarely, if ever, done.

Sometimes information that is normally outside the examiner's field of view can drift within sight. For example, if an examiner happens to look at a company's website and it states that a certain product that happens to resemble the invention being examined is for sale, then the examiner may raise the on-sale flag. Or, a disgruntled inventor who believes he or she was inadvertently omitted from a patent application may contact the patent office or file a lawsuit alleging nonjoinder of an inventor, thus putting the inventorship requirement within the examiner's field of view.

So the question arises: If at least some of the requirements for a patent fall outside the examiner's field of view, how do these requirements get verified or otherwise scrutinized before a patent issues?

The answer is simple: They don't.

Patents issue all the time without having had all the main requirements verified. That is just the nature of the examination process. The inability of the USPTO to actually confirm that all the main patent requirements are fully met prior to issuing a patent has profound consequences for the true nature of patents and how they need to be viewed in the context of running a technology business.

3.4 THE SELF-IMPOSED BUSINESS VALUE REQUIREMENT

Conspicuously absent from the main patent requirements is any kind of requisite that the invention have **business value.** The USPTO takes a free-market view of inventions. People are generally free to patent almost anything they want as long as it works like they say it works, provides at least some minimal benefit, and satisfies the requirements for a patent. This is why there are so many ridiculous inventions directed to things that people will never buy or that companies will never use. The right to life, liberty, and the pursuit of happiness as stated so exquisitely in the US Declaration of Independence clearly includes the right of a citizen (or corporation) to spend money to patent inventions that their fellow citizens (or corporations) couldn't care less about.

The proverbial mad scientist/inventor working in his or her garage with too much money and time and not enough business savvy is not the only person inventing and patenting inventions no one wants or will ever use. Technology businesses do it every day. It is up to the inventors or the technology business that is going to own the patent to have the discipline to go through the exercise of assessing whether an invention has business value and is worthy of the time, effort, and expense of a patent.

Apparently not every technology business has such discipline because a huge number of patents are filed by technology businesses that apparently want to own parts of IP space that are destined to remain unvisited. This is a dumb way to invest in real estate on earth, and it is an equally dumb way to invest in IP real estate in IP space.

A technology business needs to have a *self-imposed business value requirement* for its patents. There is some irony in the fact that business value is probably the most important requirement for patenting in a technology business and yet it needs to be self-imposed. You would think technology businesses would just do this automatically. And yet, once you have read enough patents issued to technology businesses and spent enough time observing how some technology businesses pursue their patents, you will come to the startling conclusion that this is not the case.

NOTES

1. Who wouldn't like macaroni and cheese when it's called *coquillettes avec fromage?*
2. Nuclear weapons and suicide machines, for example.
3. This rule arises through the Paris Convention for the Protection of Industrial Property (1883), a treaty relating to patent rights that most of the major countries have adopted.
4. "But I thought that the claims automatically had the effective filing date of the earlier filed provisional patent application..."

5. *Liebel-Flarsheim Co. et al. v. Medrad, Inc.,* 481 F.3d 1371, 82 USPQ2d 1687 (Fed. Cir. [S.D. Ohio] March 22, 2007).
6. In a patent infringement suit, it is first determined whether a claim is infringed. If it is, then the validity of the claim is addressed. This is because if there is no claim infringement, then there is no need to address claim validity.
7. *Therasense, Inc. et al. v. Becton, Dickinson & Co., et al.,* 649 F.3d 1276, 99 U.S.P.Q.2d 1065 (Fed.Cir.(N.D. Cal) May 25, 2011).

CHAPTER 4

PATENT PORTFOLIO THERMODYNAMICS

4.1 PATENT VALUE

One way patents are measured is by their "value." Sometimes a technology business or individuals (e.g., investors) need to know what a patent is actually worth, such as when the patent is offered for sale or for licensing. Accountants and bankers have come up with all kinds of schemes for valuing patents. These schemes all share a common feature: uncertainty. The process of valuing a patent is inherently uncertain because a patent's value depends on many factors. For example, it can be difficult to know how easily a patent might be avoided (i.e., designed around) or whether there is invalidating prior art, which if found could reduce the value of the patent to zero rather quickly. It is also difficult to know whether the patent has other flaws that could undermine whatever its value appears to be on the surface.

It is the rare patent that represents a true bottleneck, through which everyone in the technology must pass to get to their products, while also being valid.[1] The vast majority of patented inventions are incremental improvements of an existing invention and represent one of several possible ways of doing something. In fact, one approach to patent valuation is based on assessing the cost associated with having to use alternate, noninfringing versions of the invention instead of the patented version.

There are two main ways of thinking about patents and patent value. The first way looks at an individual patent, while the second way considers a collection of patents that make up a *patent portfolio.* The first way involves looking at a single patent up close, as if through a microscope. The second way involves looking at a collection of patents from far away, as if through a telescope.

If one chooses the first way and looks at a single patent, then the internal structure of that patent immediately becomes important in its valuation; that is, the entire patent, including its *file history,* needs to be scrutinized closely. Its claims must be examined in detail and compared to the market for the technology and related information, such as how long the market is expected to last, its product pricing, and so on. Assigning a value to a patent that covers an existing product with an existing market with a well-understood market trajectory and an industry standard royalty rate is a fairly straightforward exercise in accounting and thus has minimal uncertainty.

More and more technology companies, however, are creating patent portfolios to cover a much larger region of IP space than is usually possible with a single patent. Patent portfolios are created as part of a patenting strategy that relies on the perception that covering a large swath of a premium IP space is inherently valuable.

Evaluating a patent portfolio, however, is not necessarily a linear exercise of applying the single-patent valuation approach one by one to every patent in the portfolio.

This approach might work for a portfolio of just a few patents that cover products in a well-defined market. But as the number of patents increases, the single-patent valuation approach becomes less and less appropriate. There is a synergetic effect that comes with putting together a substantial collection of patents in IP space. If you think of a patent as a stake in the ground, then all the individual stakes that represent a portfolio eventually start looking like a fence. At some point, you need to start evaluating the fence and not just the fence posts. Put differently, a patent portfolio can become so large that to see how it operates as a single entity it must be viewed from afar.

4.2 PATENT PORTFOLIOS

In this IP Era of the Information Age, companies (and especially large technology businesses) can muster the resources to generate enormous patent portfolios. While portfolios can be of any size, technology businesses these days are putting together portfolios of hundreds or even thousands of patents. For example, in the summer of 2011, the bankrupt Canadian telecommunications company Nortel Networks auctioned its portfolio of about six thousand patents and patent applications. The winning bid of $4.5B went to a consortium of companies formed by Apple, EMC, Ericsson, Microsoft, Research in Motion, and Sony. The consortium outbid others such as Intel and Google. The patents and patent applications are directed to wireless technology, various Internet applications, and semiconductor technologies. A crude estimate of the value V_P of each patent in the portfolio can be found by simple math:

$$V_P = \text{total cost/no. of patents} = \$4.5 \times 10^9/6 \times 10^3 \sim \$7.5 \times 10^5$$

or about $750,000 per patent.

Also in the summer of 2011, Google announced a deal to buy Motorola Mobility Holdings, Inc. for $12.5B for the specific purpose of acquiring ownership of *approximately 17,000 patents*. Google intends to use the patents to fend off lawsuits relating to its efforts to gain access to the smartphone market. Again, the first-order valuation math is easy:

$$V_P = \$1.25 \times 10^{10}/1.7 \times 10^4 \sim \$7.35 \times 10^5$$

or about $735,000 per patent.

This value per patent is remarkably similar to the value of the Nortel patents. In fact, it would not be surprising if the Nortel IP deal were used as a comparable ("comp" in valuation lingo) for the Google IP deal.

Not to be outdone, on April 9, 2012, Microsoft announced that it had paid AOL $1B for access to AOL's portfolio of 800 or so patents by licensing the portfolio while taking outright ownership of certain of its patents. The clear implication is that Microsoft is gathering ammunition for any conflict relating to smartphones and social media technology. The valuation math for this deal is as follows:

$$V_p = \$1 \times 10^9/8 \times 10^2 \sim \$1.25 \times 10^6$$

or $1,250,000 per patent.

At this point, you might start to wonder whether it would be worth forming a company just to collect patent portfolios and then leverage them for profit. You can certainly do that, but you would not be the first. Intellectual Ventures is a prime example of a new type of IP company called a "patent aggregator" whose sole purpose is to collect portfolios of patents and leverage them. Intellectual Ventures is estimated to hold between 30,000 and 60,000 patents and is believed to be the fifth largest patent holder in the United States and the fifteenth largest worldwide.[2] The reason it is so hard to know the exact number of patents Intellectual Ventures holds is that it uses a multitude of shell companies and management entities to obscure its dealings while insulating itself from any countercharges that might arise during patent lawsuits.

4.3 THE CLASSICAL IDEAL PATENT GAS

When patents are collected into a portfolio, at some point a kind of patent thermodynamics takes over, and the portfolio is governed, based on the law of averages, by the collective behavior of all the member patents rather than by the patents as individuals. To see how this works, consider for a moment a balloon filled with a gas such as the one shown in Figure 4.1.

The kinetic theory of gases holds that the atoms in a gas have a velocity distribution, with some gas atoms moving very slowly and others zipping around very fast. Most gas atoms move at some intermediate velocity. The collective behavior of the gas atoms moving at different velocities and bouncing off the balloon wall creates pressure that defines the volume of the balloon. Note that even the atoms with a very small velocity contribute to the pressure of the gas and thus to the volume of the balloon. The atoms of the gas can be treated like little spherical particles because their internal structure is irrelevant to the calculation of the gas pressure.

The patents in a large patent portfolio can be thought of as the atoms of a patent gas in a portfolio balloon. The patents collectively give rise to a patent pressure that represents the collective value of the portfolio. As illustrated in Figure 4.2, if the portfolio is sufficiently large and if the IP space of the technology is active, then the patent atoms will typically have a value distribution that ranges from hardly any value to very valuable. Most will have some intermediate value.

One way to think of patent value is as a measure of the probability that someone might infringe it. Here, the ability of patent litigators to stretch patent claims to the limits of credulity when pressing their cases should not be underestimated. Even the patents that have hardly any value on an individual basis can contribute to the patent pressure and thus to the portfolio's value.

43

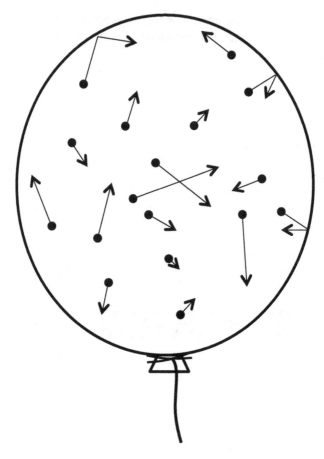

Figure 4.1 Patent portfolio balloon containing spherical patents that create patent pressure and give the patent portfolio its value.

Another point to keep in mind is that while the velocity distribution of gas atoms in the balloon generally stays the same, some atoms speed up and others slow down. The same is true for patents in a portfolio. What looks like a worthless patent today might be a valuable one in the future. As we explore in more detail later, the validity state of a given patent in the portfolio is unknown until it has been measured. It is not likely that many, if any, of the patents in a large portfolio have ever been truly measured. This uncertainty in the validity state of the patents (and patent applications) also contributes to the patent pressure. It serves as a source of angst for those looking at the portfolio and wondering how much time, effort, and expense it would take to measure even some of the patents. So the ability of patents to generate business angst in competitors is a form of business value.

At some point, the sheer number of patents in a portfolio can start to matter more than their content, especially to the folks on Wall Street. The spherical patent approximation is especially valid in cases where the patent portfolio reaches a critical mass and is treated as a single, ominous entity—a patent-based weapon of mass destruction (P-WMD).

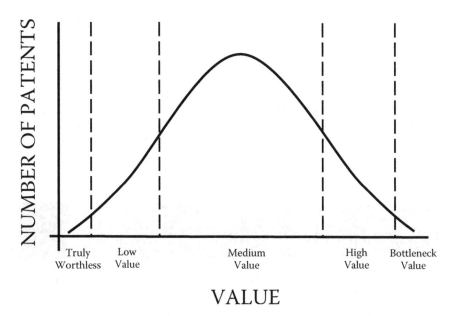

Figure 4.2 A hypothetical statistical distribution of patent value in a patent portfolio, akin to the Maxwellian velocity distribution of atoms in a balloon.

What happens to the volume of a balloon if the temperature increases or decreases? The gas pressure increases or decreases, and the balloon expands or contracts. The same thing happens to patent portfolios. If the market for the technology in the IP space heats up, then the patent pressure goes up and the value of the portfolio expands. If the market cools down, then the patent pressure goes down and the value of the portfolio shrinks.

It is interesting to note that the Nortel patent portfolio includes patent applications. Some of these applications might end up having zero value if they don't issue as patents. This highlights a critical point: The value of a patent portfolio rests in its ability to occupy a wider portion of IP space collectively than the single square covered by any one patent. It is almost certain that some of the patent applications in the Nortel portfolio and in the newly purchased Google portfolio will end up having zero value because they won't issue as patents or because the claims will have to be narrowed to the point where even intentional infringement would be a chore. On the other hand, the aforementioned angst about the scope of claims that might eventually issue from patent applications in the portfolio can be quite disconcerting—especially if they have the potential to issue with *no-shame claims,* as explained later in Chapter 7.

This uncertainty associated with patents and their valuation can shift the emphasis from individual patents to a portfolio's ability to cast a large and conspicuous enough shadow over the relevant IP space that forces others to take notice. A patent portfolio that deflects even a single patent lawsuit that threatens to burden a technology business with stiff royalty payments could very well save that business a fortune. On the other hand, a patent portfolio wielded offensively for the purpose of generating patent lawsuits can be a dangerous thing if you are on the receiving end; forcing another party to participate in an expensive and drawn-out litigation can be extremely disruptive, regardless of the outcome.

It is worth pausing here to observe how patents can act as a tax on technology. If a technology business accused of patent infringement ends up taking a license to some or all of the patents in a portfolio, then its costs increase owing to the licensing fees. If that same technology business refuses to license some or all of the patents in a portfolio, then its operating costs increase owing to the legal costs involved in defending itself in the USPTO's re-examination proceeding or in federal court. Either way, it is a cost incurred.

4.4 THE MEGAEXPENSE OF CREATING A MEGAPORTFOLIO

The strategy of creating a large patent portfolio to occupy a substantial portion of the relevant IP space can be extraordinarily expensive since by definition it involves either the filing and prosecuting of a large number of patent applications or the outright purchase of patents. Creating a megaportfolio such as the ones mentioned before that have thousands of patents costs a mega-amount of money.

If you conservatively estimate the costs of filing, prosecuting, and maintaining a US patent at $20K, then a portfolio like the Motorola Mobility Holdings, Inc. portfolio of about 17,000 patents comes in at about $340M. So the downside of the megaportfolio approach is that you need to place a very large bet that the technology will be adopted and commercially successful. This is where large technology businesses have a distinct advantage over small companies. Large technology businesses are better situated to create (or, as we have seen, purchase) patent portfolios that they can then wield with impunity while assuming the risk that the portfolio might not be as valuable as they anticipated. A patent portfolio can be made so large that, like the banks in the 2008 recession, they become "too big to fail"—meaning that, statistically speaking, the portfolio is virtually guaranteed to contain some patents capable of wreaking havoc in the particular industry. The patent balloon can become so big and potentially noxious that no one would dare try to pop it.

Like a single patent on an invention that never makes it to market, an entire patent portfolio can become worthless if its market goes away or moves in a different direction. A portfolio conceived as a patent-based WMD might very well end up as the mother of all duds. As noted earlier, the pressure of the patent portfolio balloon changes with the temperature. If the market cools off, then the patent pressure will drop. If the market goes away entirely (or if it was never really there in the first place), then the temperature drops to a point where the patent atoms can't exert enough pressure to keep the balloon from collapsing on itself.

The lifetime of a patent portfolio is affected by the lifetimes of the patents it contains.[3] Thus, playing the patent portfolio game effectively requires constantly maintaining and updating the portfolio in a manner that sustains, if not grows, the portfolio's value. For each patent atom that expires and escapes from the portfolio balloon, a new patent atom must be added to keep up the portfolio's pressure. This takes the expenditure of serious attention, hard work, and money over a sustained period. Unfortunately, most small technology businesses cannot afford to play this part of the patenting game the way the big companies do. But smaller technology businesses have other options, which we will explore in Chapter 13.

The "how hard can it be?" mentality can be dangerous when it comes to patent portfolios. It is easy to think that just because someone else struck it rich by generating an enormous patent portfolio, anyone can do it. The patent portfolios that garner all the attention in the press are the ones that hit it big, like the smartphone patent portfolios. No one is going to write news articles about the large patent portfolio that no one bought or that is worth less than what it cost to produce. By dumb luck alone, a certain percentage of all patent portfolios in existence at any one time are going to end up having extraordinary value. If one were to take all the patent portfolios in existence today, they would likely have a value distribution akin to that of Figure 4.2, with most portfolios having moderate value, some having essentially no value, and a few having enormous value. The randomness of life, the vicissitudes of markets, and the hidden role of chance are all factors that determine whether a given patent portfolio is going to prove to be a successful patent-based WMD or a dud (Taleb 2004).

NOTES

1. This is not to be confused with patents that appear to be "bottlenecks" only because they have ridiculously overbroad claims. We discuss such patents in Chapter 7.
2. Ewing, T., and R. Feldman. 2012. The giants among us. *Stanford Technology Law Review* 1.
3. Generally speaking and assuming a spherical patent, patents in the United States have a life span of 20 years as measured from the earliest effective filing date.

REFERENCE

Taleb, N.N. 2004. *Fooled by Randomness: The Hidden Role of Chance in Life and in the Markets*. New York: Random House.

CHAPTER 5

CLASSICAL AND QUANTUM PATENT MECHANICS

5.1 CLASSICAL PATENT MECHANICS

The branch of physics referred to as *classical mechanics* describes and predicts with great accuracy the motion of macroscopic objects moving much slower than the speed of light when subject to one or more forces. At the center of classical mechanics is Newton's famous second law of motion: force = mass × acceleration. Classical mechanics is deterministic in that it exactly describes (within its limitations) the future motion of an object, provided that the object's mass, initial position, and velocity, as well as the forces acting on the object, are sufficiently well known. The objects in classical mechanics are often approximated as fictional particles of a certain mass, their sizes being negligible or irrelevant. In cases where the physical extent of an object needs to come into play, the object often takes the form of a sphere or some other calculation-friendly shape (recall, for example, our *spherical asteroid* from the preface to this book).

A classical mechanics view of the world is attractive and comforting. Its inherent determinism allows us to predict with "certainty" what is going to happen to any physical system, be it billiard balls rolling around a table, a kicked soccer ball, a moon rocket, planets orbiting a star, or galaxies colliding. It provides a framework for thinking about and understanding the universe at the macroscopic scales that humans can grasp, and it is how almost everyone on the planet views the world.

For these same reasons, most technology workers view and understand patents in a linear and deterministic way that I call *classical patent mechanics.* This view of patents relies on a host of simplifying assumptions, including the perfectly spherical patent—a single uniform entity with no internal structure and with a single property of "value" akin to mass. Everything about patents is clear and predictable, as long as you know all the forces involved. It's not really all that complicated.

5.2 A BRIEF CLASSICAL PATENT MECHANICS VIEW OF THE PATENTING PROCESS

From the viewpoint of classical patent mechanics, a patent application filed in the USPTO has a clear and predictable trajectory. The patent application is subject only to the known forces of the patent laws that the patent examiner applies to arrive at one of only two possible outcomes: Either the invention is patentable and a patent issues or the invention is unpatentable and a patent fails to issue. If the invention is patentable, then a patent issues in a definite state of being valid and enforceable. The patent's claims, which describe the legal scope of the invention in the aforementioned "patentese," have sharp boundaries based on their plain meaning.

Under the first-to-file system, whoever files a patent application on the invention first wins the race and will be entitled to the patent if the invention is indeed patentable. The issued patent, like all issued patents, has a definite life span of exactly 20 years from its filing date. After that, the patent expires and the invention becomes available to the public. A patent has value if there is some definite market need for the invention; otherwise, it has no value. If the patent is infringed (i.e., used, made, sold, offered for sale, or imported into the United States without the patent owner's permission), then the patent owner has three basic choices:

- Do nothing
- License the invention
- Not license the invention and get an injunction (i.e., court order) to prevent the infringer from continuing the infringing activity or activities[1]

Fortunately, under classical patent mechanics, an infringer can't sue the patent owner or try to invalidate the patent unless the patent owner goes after the infringer first. The value of a patent portfolio is simply the sum of the value of each patent in the portfolio.

This, in a nutshell, is what patents are all about. What's the big deal? How hard can it be?

5.3 A CLOSER (BUT STILL BRIEF) LOOK AT THE PATENTING PROCESS

The preceding brief description of the patenting process according to classical patent mechanics sounds pretty reasonable and straightforward. The problem is that *it is only about half right.* It omits important details that can lead, at best, to inaccurate conclusions and, at worst, to unmitigated disaster.

For example, as we shall soon see, the linear and eminently reasonable assumption that the patent office issues only valid patents is false. Invalid patents can and do issue from the patent office and can wreak havoc on a technology business.

The assumption that an issued patent is valid and enforceable is not quite right either, especially in view of the fact that the USPTO sometimes inadvertently issues invalid patents. This fact is for many a huge source of angst because this phenomenon is inexplicable using classical patent mechanics. "Why can't they just get it right?" is the typical lament of those who ascribe to the classical patent mechanics viewpoint. The reality is that an issued patent is only *presumed* valid and enforceable. This presumption assumes that all the main requirements for a patent have been met, even those that are typically outside the examiner's field of view. Furthermore, the presumption of validity may be rebutted (i.e., overcome) through re-examinations and postgrant proceedings at the USPTO or through the federal court system if a strong enough case can be made.

One of the more interesting circumstances under which the validity of a patent can be challenged is when the patent owner invites another party to consider licensing the patent. Even if the invitation is purposely worded to be friendly and nonthreatening, under current law it can serve as the basis for the potential licensee to initiate a declaratory judgment suit against the patent owner to challenge the validity of the patent in question.[2] So a potential infringer does not have to wait to be sued in order to sue the patent owner and seek the invalidation of the patent in court. Put differently, *a patent owner can be dragged into a costly patent lawsuit merely by trying to license the patent.* The defendant in a patent infringement lawsuit will almost always challenge patent validity for the very sensible reason that such challenges are often successful.

Likewise, the assumption that the claims have sharp boundaries based on plain meaning is only sometimes true. The scope of what a claim actually covers needs to be construed based on the language used in the description of the invention, the file history, and the common meaning of the terms used in the particular technology or art. The scope of coverage that a claim provides based on a straightforward reading can differ substantially from the coverage it provides based on the construed meaning of certain claim terms.

For example, it may be that a claim uses the term "fluid," whose plain meaning includes both a liquid and a gas. However, the description of the invention or the file history may limit the meaning of the term "fluid" to a liquid. So reading a patent claim without reading the entire patent and its file history can lead to an erroneous understanding of the claim's actual scope of coverage. And an erroneous understanding of the scope of a claim can translate into the loss (or gain) of untold millions of dollars.

A claim is said to be *literally infringed* when the accused product or method falls within the construed scope of the claim. However, there are situations where the accused product or method can fall outside the construed scope of the claim and still infringe the claim. This can happen through the application of what is called the **doctrine of equivalents,** which is discussed in Chapter 7. So where classical patent mechanics always sees a bright-line boundary for claims that clearly separates infringement from noninfringement, the actual boundary can be quite fuzzy.

Under the US version of the first-to-file system, the first filer is not always the winner. For example, if the first filer didn't actually invent the invention but stole it from the real inventor, the first filer's application won't count. Moreover, under a scenario we discuss in Chapter 6, a person who is first to invent an invention and who is also the first to file a patent application on the invention can still lose out to a second filer who subsequently files a patent application on the very same invention.

The assumption that a patent has a 20-year term as measured from its filing date is not entirely true either. First, the term is measured from its effective filing date, which can be the application's filing date in another country and not the filing date for the US application. Under certain circumstances, the patent term can be extended—for example, if the patent application was pending in the USPTO for more than 3 years.[3] Likewise, in certain cases a patent may be subject to a terminal disclaimer that disclaims a portion of the 20-year patent term. If maintenance fees are not paid during the third, seventh, and eleventh years after issuance, the patent will "expire" (lapse) before its full term is completed—maybe.

At an even finer level of granularity is the rule that allows a patent owner to petition the USPTO to reinstate a lapsed patent provided the owner makes a delayed payment of the maintenance fee.[4] So you cannot assume that a lapsed patent has really lapsed. At the microscopic level, in some situations the timely payment of maintenance fees might not be sufficient to prevent a patent from lapsing. It turns out that the USPTO recognizes patent owners as either small entities or large entities, with the small entities paying smaller fees. If a large entity pays the small entity maintenance fees, its patent could be declared unenforceable.[5] So much for the simple spherical patent view of the fixed and definite 20-year patent term.

The assumption that patent value is based on a definite market need for the invention is also overly simplistic in view of the emerging role of patent portfolios. As we saw in Chapter 4, how we view the value of a patent depends on our frame of reference. The value of a single patent considered by itself can be very different from its value when considered en masse with a bunch of other patents in a patent portfolio. So, for a large enough portfolio, the value can be more than the sum of its parts.

Finally, the assumption that a patent owner can obtain an injunction (court order) to stop an alleged infringer who refuses to take a license and keeps on infringing relies on the ready availability of such an injunction. In fact, there is no guarantee that an injunction can be obtained, and a fairly recent court case[6] has made getting injunctions for patent infringement more difficult than ever. This is an example of how court decisions can have a huge impact on patenting. Patent owners' inability to get what were once essentially automatic injunctions against an alleged infringer has seriously diminished their leverage.

The point of this discussion is that classical patent mechanics falls well short of providing an accurate model for what patents really are, how they are obtained, and how they actually work. It does not account for nuances that almost always matter. This, by the way, goes a long way in explaining why one can rarely get a straight answer from a patent attorney to a simple question about patents. The answer almost always starts with the phrase: "It depends…." Answers to patent-related questions are usually fact intensive, not to mention law intensive, rule intensive, and regulation intensive. A patent attorney therefore usually needs to gather a litany of initial conditions and boundary conditions before he or she can begin to formulate an answer. While to some extent this speaks to how uptight some patent attorneys can be, to a greater extent is speaks to the fact that patents and patent-related issues tend to be far more complex than most people think they are.

Because classical patent mechanics fails to describe patents and their behavior accurately, a more sophisticated view is needed that better explains their mysterious properties.

5.4 QUANTUM MECHANICS INTERLUDE

The inability of classical mechanics to describe phenomena at microscopic scales accurately is a consequence of the certitude and definiteness associated with its deterministic view. The development of **quantum mechanics** to describe the microscopic world of atoms and subatomic particles accurately required a transition to a stochastic (i.e., nondeterministic) view based on uncertainty and probabilities.

For example, where classical mechanics says that an electron or other subatomic particle could have its position and velocity exactly determined, the Heisenberg uncertainty principle of quantum mechanics places a limit on how well we can measure the position and velocity of a subatomic particle at the same time. In lay terms, this means that a moving subatomic particle cannot be precisely located at any given time. One is left with talking about where the subatomic particle is *likely* to be. Thus, the solar-system model of an atom, in which electrons have well-defined orbits around the atom's nucleus, gives way to the more sophisticated and realistic model wherein the orbits are represented by *probability distributions* that describe where electrons are most likely to be found.

The stochastic quality of quantum mechanics also means that a subatomic particle is not in a definite state but rather in a *superposition of all of its possible states* until it interacts with an observer or is otherwise measured.

To get a better handle on this superposition business (which we will employ momentarily in the patent context), let's consider the situation of Figure 5.1(a), which shows a person flipping a coin into a box and closing the lid before the coin's final state (heads, H, or tails, T) can be viewed.

The coin represents a "coin system" that has two possible states: heads or tails. Under the classical mechanics viewpoint shown in Figure 5.1(b), the coin inside the box will definitely be in either one of its two possible states, even though it can't be seen. Opening the box to see the coin will reveal what state the coin system was in once the coin settled down inside the box.

Under the quantum mechanics viewpoint shown in Figure 5.1(c), the coin inside the box is not in one state or the other. Rather, it is in a superposition of its possible states—namely, heads and tails. It is not in any one state of heads or tails. The act of observing the coin—that is, the measurement of the coin's state—causes the coin to take on one of its possible states. In quantum-mechanical terms, the coin has a *wavefunction* that describes the superposition of all the coin's possible states. The coin's wavefunction collapses into one of the possible states only upon being measured. In the case of the coin system, the chance of the coin's wavefunction collapsing to heads or tails is 50:50.

The key principle to take away is that, from a quantum mechanics viewpoint, the act of making a measurement of a system affects the outcome of the system, and until the measurement is made, there is not yet an outcome. No one can say what the state of the system actually is until it is measured. While this notion of the superposition of states does not actually apply to macroscopic objects such

Flip the coin into the box and immediately close the lid

(a)

CLASSICAL VIEW

(b)

Figure 5.1 (a–c) Classical versus quantum views of the uncertainty in the state of an unmeasured object using a two-state coin system.

QUANTUM VIEW

(c)

Figure 5.1 (*Continued*)

as coins, it does apply to subatomic objects, which do not have an analogous macroscopic structure to which we can relate. Apparently, objects like atoms, electrons, protons, photons, and molecules have unique properties that simply cannot be explained by analogy to the macroscopic world.

The relevance of this detour through quantum mechanics will become clear momentarily.

5.5 QUANTUM PATENT MECHANICS AND THE VALIDITY UNCERTAINTY PRINCIPLE

As it turns out, patents are like microscopic particles in that they are rife with uncertainty. How so? Well, in Chapter 3 we identified eleven main patent require-ments. Of these eleven requirements, we saw in Table 3.2 that only six of them fell within the examiner's field of view. That table is included below by way of reminder.

Even the evaluation of the six requirements that are within the examiner's field of view and are usually directly examined is subject to substantial error. Assessing novelty and nonobviousness requires comparing the claimed inven-tion to the prior art. But, as we noted earlier, it may be that the examiner does not have the most relevant prior art in front of him or her. Also, the obviousness

Table 3.2 The examiner field of view

Within the field of view	Outside the field of view
• Statutory subject matter	• No public use
• Novelty	• No sale
• Nonobviousness	• Inventorship
• Utility	• No inequitable conduct
• Enablement	• Effective filing date
• Written description	

analysis is complicated and subjective, and the examiner might not get it right.[7] Making the right call on the enablement and written description requirements hinges on a careful reading of the entire specification, which is a tedious and time-consuming task that is not always performed as part of the examination. There is thus substantial uncertainty even for those requirements within the examiner's field of view.

The collective uncertainty associated with the patent examination and issuance process and the numerous and complex rules that govern patents mean that patents are governed by quantum patent mechanics rather than by classical patent mechanics.

One of the most important rules in quantum patent mechanics is the *validity uncertainty principle,* which can be stated as follows:

> **A patent exists in a superposition of a valid state and an invalid state until its validity is measured.**

This is to say, the validity of a patent is **uncertain** until it is **measured.** The *presumption of validity* only matters here in the context of who has the burden of proof to show that a patent is valid—namely, the patent challenger. We discuss later what is meant by "measured." But first, there is one more twist to this quantum patent mechanics view: Not all patents are created equal. The probability that a given patent's wavefunction will collapse to either a valid or an invalid state is not just a 50:50 coin toss. Instead, each patent has a unique wavefunction that depends strongly on how well the patent application was prepared, how completely it attempted to satisfy all the patent requirements, and whether the patent application received a proper examination in the USPTO. These variables speak to the notion of **patent quality,** which we address in more detail in Chapter 7.

5.6 TO THE USPTO, AUNT BETTY AND INTEL ARE EQUAL

At this point you might be wondering to yourself, "Why can't the USPTO just rigorously measure patent validity during the examination?" This is a good question. Essentially, the USPTO's bureaucratic nature combined with the nature and scope of the examination task makes it inherently unable to perform such a measurement. To be able to measure all patent applications with sufficient rigor to collapse the patent's wavefunction to a definite state would require an examination process far beyond what the government and patent applicants could afford. The USPTO's budget would need to be more like the Pentagon's budget. It would also take way too much time and consume far too many resources. Furthermore, there are no free-market forces at work since the USPTO has a monopoly on patent examination. As a consequence, the patent examination process at the USPTO is going to have substantial entropy. All the USPTO can do is shape the patent wavefunction.

In fact, it already takes years to get a patent application examined. We can't have patent examiners acting like FBI agents and running down leads regarding the possible public use of Aunt Betty's gas-powered tea-cozy invention. Now this might be a useful exercise in the case of a new design from Intel for a logic chip that has profound implications for computing and the potential for huge economic success. But guess what—it's a free market for inventions. The USPTO isn't set up to make marketability and value judgments about inventions made by sophisticated and economically important technology businesses versus those made by Aunt Betty. And who is to say that Aunt Betty might not be able to license her invention at a profit?

5.7 MEASURING VERSUS FILTERING

Figures 5.2(a) and 5.2(b) contrast the classical and quantum views of how the USPTO assesses whether a patent application meets all the requirements to issue as a patent. As shown in Figure 5.2(a), under the classical patent mechanics viewpoint, patent applications are filed in the USPTO. The USPTO machinery has an array of requirement detectors that detect whether the patent application meets the requirements. If one of the detectors indicates that the patent application has failed to meet the corresponding requirement, then the patent application gets rejected. The patent owner can then revise the application so that it will pass all the measurements made by the requirement detectors. At that point, the USPTO will consider the invention patentable and issue a valid patent.

With reference now to Figure 5.2(b), under the quantum patent mechanics viewpoint, patent applications are filed in the USPTO. Each patent application exists in a superposition of valid and invalid states because its validity has not yet been measured. The applications are then processed by the USPTO.

However, the USPTO does not actually measure the patent applications with any requirement detectors. Rather, the applications are passed through requirement *filters,* like using a sifter to sift rocks. Some patent applications make it through all the filters and become issued patents while others get blocked (rejected) by at least one of the filters. If the patent application gets rejected, then the patent owner can revise it so that it passes through the problematic filter. There are no detectors measuring anything. The requirement filters, at best, cause the patent application to be transformed along the way so that it can pass through all the filters.

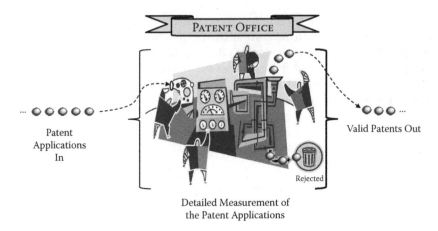

Figure 5.2 (a) The classical patent mechanics view of the USPTO.

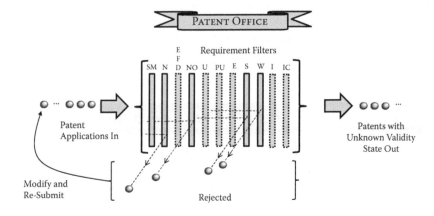

Figure 5.2 (b) The quantum patent mechanics view of the USPTO.

Now, with a perfect sifter, you can obtain something like a measurement because at least you can say that the rocks that make it through are of a given size or smaller and the ones rejected are of a given size or larger. However, if the sifter has some uncertainty, such as holes of varying size, including some gaping holes, then the sifter becomes dysfunctional and you cannot count on it to sift rocks with accuracy.

As we discussed before, each requirement contains some uncertainty, and this uncertainty makes each requirement filter dysfunctional to some degree. So now we are back to talking about the probability and statistics associated with each filter since there is now a chance that a given filter can malfunction and pass a patent application that it should not pass or block a patent application it should not block.

Now here is an interesting thing to think about: If a filter has a *nonzero probability* of passing a patent application that it shouldn't pass, then you can keep trying to send the same (or essentially the same) patent application through the *same* filter until the odds catch up with it and it finally passes through.[8]

The five filters we identified as falling outside the patent examiner's field of view can be so dysfunctional as to reach the point of nonfunction. These filters are usually all but open windows. They are essentially nonfilters. They will pass almost every patent application. The effective filing date filter often does not have the required resolution to spot effective filing date issues, so it can be essentially an open window (think about the thinly enabled provisional patent application that never gets scrutiny). In fact, the effective filing date filter won't ever stop a patent from issuing; it just won't get the effective filing date right so that, down the road, it can be easier for someone else to challenge the claims (we discuss this scenario in more detail in Chapter 6).

The utility filter is also within the examiner's field of view and has very little dysfunction. But the utility filter is enormously wide, passing almost any invention except those that are inherently inoperable. The upshot is that the requirement filters for public use, on sale, inventorship, no inequitable conduct, effective filing date, and utility just aren't very good. They are illustrated in phantom in Figure 5.2(b) because they are not quite there.

The remaining four filters for novelty, nonobvious, enablement, and written description usually end up doing all of the heavy filtering. But even if these workhorse filters were perfect (and they are not), the dysfunctionality of the other filters introduces substantial uncertainty about whether an issued patent is actually valid.

5.8 LET'S PARTE!

It is important to understand that the patent examination process is not adversarial. There is the patent applicant and there is the patent examiner who plays the part of a neutral minijudge. There is no opposing party like there is in a courtroom. In legal jargon, the patent examination process is *ex parte,* or "one-sided."

If you want to see some serious pushback in the patent examination process and a high level of enthusiasm for ensuring that all the patent requirement filters are working to their maximum capacity, add to the process an adverse party who stands to lose a substantial sum of money if the patent issues.

So the ex parte nature of the patent application examination process prevents the USPTO from making the kind of measurement most people think it actually makes. When a patent application is filed, there is no "other party" that stands ready to oppose the application. That can only happen later on in the process, after the patent issues.

A recent change in the patent laws introduced by the AIA allows post-issuance challenges to patents in what is called an *inter partes review.* This is a two-sided review of a patent that takes place under the aegis of the USPTO rather than in federal court. The inter partes review through the USPTO is much like the opposition proceedings used to oppose issued European patents.

But note the key point: The patent examination process itself remains one sided until the patent issues. It is only after the fact that an opposing party can step in and seek an inter partes review of the issued patent. The level of scrutiny that a patent application normally receives during its ex parte examination at the USPTO is "good enough for government work." But this is not good enough to establish the validity state of the issued patent for certain. A serious measurement of the patent's validity happens only when someone else with enough money at stake feels compelled to make it happen.

5.9 THE PRESUMPTION OF VALIDITY

As discussed previously, an issued patent is entitled to a presumption of validity. This presumption places the burden of proving invalidity squarely on the shoulders of the one challenging the patent. In the meantime, however, this presumption can create the false impression that the validity of issued patents is strong and that a given patent's wavefunction is more likely to collapse to a valid state than to an invalid state if the patent were actually to be measured.

The true strength of this presumption of validity, however, is a function of the particular patent in question. If a patent's wavefunction has a high likelihood of collapsing to an invalid state owing, say, to a truckload of prior art that clearly anticipates the invention, then the barrier posed by the presumption of validity will look something like a well-worn speed bump. If, on the other hand, a patent's wavefunction has a high likelihood of collapsing to a valid state, then the presumption of validity barrier can loom large as Mount Everest. It all depends on the particular patent being measured.

All the presumption of validity means is that someone who wishes to challenge the validity of a patent in court has the burden of proof to establish invalidity to a *clear and convincing* standard of evidence. This standard corresponds to establishing the "truth" of the matter in question to a reasonable degree of certainty. This reasonable degree of certainty lies somewhere above the preponderance of evidence standard (greater than 50% convincing) and below the beyond a reasonable doubt standard (greater than about 95% convincing). This level of proof is not impossible to meet, especially if the patent is truly weak. If sufficient evidence is presented to overcome (i.e., rebut) the presumption of validity, then the burden shifts to the patent owner to show that the patent is in fact valid. So the shifting of the burden of proof with respect to patent validity is going to depend upon the particular patent being scrutinized.

In *KSR v. Teleflex,* a Supreme Court case we discuss in detail in Chapter 10, the Court was asked to opine on whether the presumption of a patent's validity was voided if a challenger presented important ("material") prior art that was not before the examiner during the examination. While the court did not answer this question directly, in the *KSR* case this was exactly the situation. The patent in question (the Engelgau patent) was examined without a key piece of prior art (the Asano patent). The Court stated:

> We need not reach the question whether the failure to disclose Asano
> during the prosecution of Engelgau voids the presumption of validity
> given to issued patents, for claim 4 is obvious despite the presump-
> tion. We nevertheless think it appropriate to note that the rationale

underlying the presumption that the PTO, in its expertise, has approved the claim seems much diminished here.

The Court here is taking a commonsense view of matters and notes that even if there is a presumption of validity, this presumption may be easier to overturn if it can be proved that the examination was not conducted with all of the most relevant prior art at hand.

5.10 IF A PATENT ISSUES IN A FOREST...

If a patent issues in a forest and no one reads it, is it valid?

Most patents are never scrutinized for validity after they issue and live their entire lives in a superposition of valid and invalid states. A patent that has no perceived value to anyone will generate no interest in its validity state and will never be measured.

No one is going to take the trouble to investigate a patent's validity just for fun. Validity is first investigated only when there is sufficient financial motivation to do so. A validity investigation requires the services of a patent attorney who is qualified to render a legal opinion. One financial motivation for initiating such an investigation is a patent infringement dispute where the accused infringer needs to invalidate the patent (or at least the problematic claims) to avoid paying a huge sum of money in damages to the patent owner. Another motivation is in connection with a licensing deal, where the potential licensee wants a better understanding of how good a particular patent is in order to assess how much, if anything, to pay. A licensor might also seek to assess validity in order to see how much his or her patent is actually worth and to get out in front of any problematic validity issues the patent might have.

To prepare a legal opinion on patent validity properly, a patent attorney must perform a great deal of groundwork, turning over all the stones and examining every minute detail to get to the bottom of the circumstances surrounding the examination and issuance of the patent. The entire patent document and its file history are scrutinized from a legal viewpoint to see if all the requirements for the patent have been met. Patent validity opinions can cost anywhere from about $5,000 in a very simple case to upward of $100,000 in situations where the gloves have come off and someone wants to scorch the earth to find every piece of prior art and track down every possible lead regarding public use, on-sale activity, and every other patent requirement.

But at the end of the day, a legal opinion is just that: an *opinion*—not the last word. All it can do is provide information about the patent's wavefunction for the purpose of better understanding its chance of collapsing to a valid or invalid state if and when the patent is *really* measured. No patent attorney worth his or her salt (or who values his or her malpractice insurance policy) will write a legal opinion and guarantee that it is 100% accurate. That means that the opinion, like the USPTO's examination, contains some uncertainty, which in turn means that it can't collapse the patent's wavefunction.

A legal opinion from an attorney is therefore not a measurement of patent validity; it is only an analysis of the patent's wavefunction that seeks to predict which way the wavefunction will collapse when it is really measured.

So how is patent validity really measured?

5.11 MEASURING PATENT VALIDITY

We observed before that a proper measurement of a patent application requires an inter partes proceeding, where each side is motivated to win. The stakes now are much higher, and the levels of enthusiasm and tension involved are much greater than in the ex parte proceeding. If the patent examination process resembles a dance, then an inter partes proceeding in federal court is more like a mixed martial arts competition. It is confrontational and aggressive, and can get rather nasty.

Figure 5.3 illustrates the two main routes for conducting an inter partes measurement of a patent to establish its validity. The first route is the aforementioned inter partes review conducted through the USPTO. This review is essentially an ersatz litigation conducted in front of the Patent Trial and Appeal Board (PTAB), which is a panel at the USPTO made up of three administrative-law judges. One may obtain an inter partes review by filing the appropriate petition and fees in the USPTO. The petition will be granted only if it is determined that there is a reasonable likelihood that the petitioner would prevail with respect to at least one of the claims challenged in the petition.

Just getting the petition granted would seem to forebode a positive outcome for the petitioner. In conducting the inter partes review, the USPTO is authorized to institute litigation-type discovery rules, which are rules that govern how each side can obtain information from the other during litigation. The petition for an

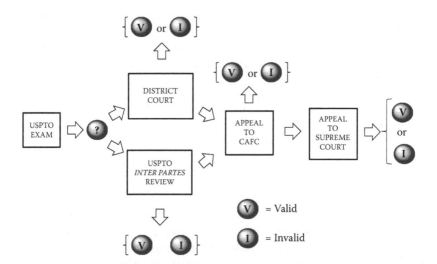

Figure 5.3 The federal court system as the ultimate measurer of patent validity.

inter partes review of an issued patent cannot be filed until 9 months after the patent has been granted or until the date of termination of a postgrant review, whichever is later. Only granted patents and printed publications can be used in the review—a far more limited scope than what can be used in challenging a patent in federal court.

If you choose the inter partes review route, then, as the petitioner, you are relinquishing your right to later assert other grounds for invalidity. An appeal of an inter partes review bypasses the federal district court and goes straight to the Court of Appeals for the Federal Circuit (CAFC), which reviews the case based solely on the record in the USPTO. No new evidence can be introduced. So taking the inter partes review route through the USPTO definitely has its limitations.

The second route for measuring a patent's validity is pursuing litigation in the federal court system. Since patent law is federal law, the federal court system handles patent cases, including questions of patent validity. The district courts are the lowest federal courts that handle patent cases. Appeals from a district court go to the CAFC. A fair number of patent disputes begin and end at the district court level. When they do, the district court's decision about whether the patent is valid or invalid is the last word. The measurement is done. The patent's wavefunction has been collapsed. End of story.

However, some district court patent decisions are appealed to the CAFC. While the district courts are decent measurers of patent validity, the CAFC is a very good measure of patent validity. Moreover, the CAFC will review the district court case "de novo," meaning that it will not limit itself to the record of the district court. Consequently, whatever the CAFC decides about patent validity is usually the last word on the matter. Note that if the CAFC reverses the district-court validity measurement (which happens quite frequently), then the district-court measurement was only illusory.

The only other remaining measurer of patent validity is the US Supreme Court, which is the ultimate and absolute measurer of patent validity (and everything else legal), so whatever it says about a patent's validity is truly the last word on the subject. However, the US Supreme Court is generally not interested in making patent validity measurements. So the CAFC is, for all practical purposes, the last resort for such a measurement, which by definition makes it an excellent measurer of validity. There are some statistics that indicate that, at the CAFC level, the ratio of patents found valid to those found invalid is about 60:40.[9]

The actual measurement of a patent by a court is a relatively rare event given the number of issued patents that are enforceable at any given time. For example, according to government statistics, in 2010 there were 3,260 patent lawsuits filed. During 2010, there were (by rough approximation) about three million unexpired patents. Of all the possible patents available for litigation in 2010, only about one out of every thousand patents reached the federal court for a measurement. And even then, many of the lawsuits never reached a determination of validity because the court case never went that far (e.g., the case was settled out of court).

5.12 WHAT, THEN, DOES THE PATENT OFFICE MEASURE?

If the patent office provides only a filtering function for a patent's **substantive** requirements, does it actually ever measure anything? The answer is yes. It is actually very good at measuring the innumerable **procedural** requirements for obtaining a patent.

When the patent office receives a patent application, it first scrutinizes the application to make sure it complies with the numerous procedural requirements that make it easier for the patent office to examine the onslaught of patent applications it receives on a daily basis (about 1,500 **per day** on average in 2012). If filers made up their own rules about what could be included in a patent application and about its format, writing style, types of figures, font, and so on, the patent applications flowing in would not be amenable to any kind of uniform processing. All patent applications must look the same and have the same defined parts if they are to be subject to examination using a set of unified (if not arcane) rules.

If you submit a patent application whose abstract exceeds 150 words, then there is an excellent chance that the USPTO will measure the word count and object to the abstract, since the *Manual of Patent Examining Procedure* (MPEP), in section 608.01(b), states that the abstract must have 150 words or fewer. If a claim does not properly introduce a claim element before referring to the element with the word "the" (as in "the widget"), the claim will almost certainly be objected to for "a lack of antecedent basis" for the referred-to item. If a reference number is missing in a drawing, or if the lines are not clear, or if the font size of the reference numbers is not within the required size range (or if any one of a hundred other potential problems you can have with a patent drawing occurs), then the drawing will be objected to. If the margins of the specification are out of whack or the font size is wrong, then correction will be required. If, when amending the claims in an office action, you leave out even one of the status identifiers for the claim (e.g., "currently amended," "original," etc.), then the entire reply will likely be rejected and you will have to resubmit the amendment all over again, this time including the one status identifier that was missing. If you try to amend the application by adding new information, then the examiner will very likely reject the amendment as adding "new matter." If you leave out the period at the end of a claim sentence, then the claim will very likely be objected to.

The level of scrutiny that the patent office applies to patent applications at the procedural level is enviable given the size of the bureaucracy and the numbers of applications it needs to process.

NOTES

1. An injunction is a court order directed either to stopping one party from taking certain actions or to compelling them to take certain actions.

2. This circumstance often gives rise to another infamous IP BITT: "BITT…we could just send a letter to the infringer to scare the company into taking a license. What is the harm in asking?"
3. The conditions for patent-term adjustment are set forth in 35 USC §154(b).
4. 37 CFR §1.378.
5. *Unenforceable* is different from *invalid;* the former usually refers to inequitable conduct that results in the patent rights being taken away and the latter involves a substantive legal flaw—for example, lack of novelty.
6. *eBay Inc. v. MercExchange, L.L.C.,* 547 U.S. 388 (2006).
7. The nadir of the dysfunctionality of the novelty and nonobvious filters to date is arguably US Patent No. 6,368,227, entitled "Method of swinging on a swing."
8. Consider, for example, Apple's US Patent No. 8,086,604, entitled "Universal interface for retrieval of information in a computer system," which finally made it through all the requirement filters on its tenth try with only some minor tweaking. This patent was then leveraged by Apple against Samsung's smartphones and tablets.
9. See, for example, Allison, J. R., and M. A. Lemley. 1998. Empirical evidence on the validity of litigated patents. *AIPLA Quarterly Journal* 26:185: (54% valid/46% invalid); Dunner, D., J. Jakes, and J. Karceski. 1995. A statistical look at the federal circuit's patent decisions: 1982–1994. 5 *FED. CIR. B. J.* 151 (1995): 153–155 (58% valid/42% invalid).

CHAPTER 6

PROVISIONAL PATENT APPLICATIONS REVEALED

6.1 INTRODUCTION

Obtaining a US patent on an invention requires filing a patent application in the USPTO so that it can be examined. Most patent applications are "utility" patent applications, which cover the kinds of inventions having some functional use (as opposed to a "design" patent application, which covers ornamental aspects of articles of manufacture), and are the kind with which most people are familiar. Utility patent applications fall into five main categories, namely:

- Mechanical
- Electrical
- Chemical
- Pharmaceutical
- Software

Patents that issue from utility patent applications are called, as you might expect, **utility patents.**

There are two flavors of utility patent applications: provisional and nonprovisional. This chapter is devoted entirely to **provisional patent applications,** which are profoundly misunderstood by many and too often used incorrectly by people working in technology businesses.

The structural and formatting requirements of a provisional patent application are much less formal than those of a nonprovisional (or "regular") patent application. While the USPTO prefers that a provisional patent application be structured like a nonprovisional patent application, it's not required. A provisional patent application need not, for example, include claims. The drawings can be informal and need not be black and white or satisfy a host of tedious requirements. A provisional patent application is not examined. It is given only a cursory inspection to make sure that it has a proper cover sheet, that the required fee has been paid, and that the document itself is a reasonable description of the invention. By all accounts, this last inspection point is cursory in the extreme.

Consequently, unless the provisional patent application is written on paper towels or chiseled into chunks of slate, it is going to be accepted by the USPTO. It will be accorded a serial number and a filing date so long as it is filed with the proper provisional application cover sheet and the nominal fee is paid (and this fee can be paid after filing).

Because of their informality, provisional applications are easier than nonprovisional applications to prepare and file. Also, the filing fee is a fraction of the cost of an nonprovisional patent application. A provisional patent application never issues as a patent. This, by the way, means that there is no such thing as a "provisional patent." So when someone at the IP bazaar accosts you and offers to prepare a "provisional patent" for you, run the other way. A provisional patent application is merely a placeholder application that serves to secure an effective filing date. It cannot claim an effective filing date from another patent application that was filed earlier, either in the United States or elsewhere. A patent or patent application that cites an application filed earlier to establish an effective filing date is said to "claim priority" from the application filed earlier.

6.2 MORE BASIC INFORMATION

A provisional patent application expires 1 year after its filing date. Accordingly, a nonprovisional patent application that intends to claim priority from a provisional patent application must be filed within the 1-year deadline and must expressly claim priority from the provisional application. Likewise, any non-US patent applications that seek to claim priority from the US provisional patent application need to be filed within the 1-year deadline.

A provisional patent application is merely a stake in the ground intended to establish the effective filing date for an invention. It is almost always preferable to establish the earliest possible effective filing date because prior art is, by definition, all information that exists prior to the effective filing date (with some exceptions). Also, properly establishing an early effective filing date can allow a disclosure of the invention by the inventor without acting as a bar to the patentability of the invention.

For example, if an invention is publicly disclosed by the inventor (say, by a commercial use or a sale, or a published journal article), the inventor can file a provisional patent application within 1 year of the first public disclosure. A nonprovisional patent application can then be filed within 1 year of the provisional patent application, keeping the patent application process moving forward while maintaining the effective filing date as established by the provisional application filing.

6.3 MORE ABOUT THE EFFECTIVE FILING DATE

Now, as you will recall, the proper establishment of an effective filing date was identified in Chapter 3 as one of the main requirements for a patent. There are two reasons for this. First, if the earliest effective filing date you are seeking for the patent application is not properly established, then you are not entitled to it. Second, most IP spaces associated with technology businesses are already crowded, and an effective filing date that precedes that of a competitor's patent application by even 1 day can make all the difference. There is usually a great deal of competition among different technology businesses, all of which are working hard to develop and patent very similar inventions at nearly the same time within a given sector of the market.

The relatively recent change in the US patenting system to a first-to-file system (assuming a spherical patent) makes a patent application's effective filing date all the more important. There is little room for dawdling when it comes to developing and patenting technology inventions in a first-to-file patenting system. The technology business that strolls to the patent office to file its patent applications and secure its effective filing date is going to be bowled over by the technology businesses that are running to the patent office.

Once filed, the provisional patent application is docketed and filed away in the USPTO. If a follow-on nonprovisional patent application is filed that uses the right words and the right paperwork to claim priority from the provisional patent application, then it will generally be granted an effective filing date that is the date of the provisional patent application. If a patent issues from the nonprovisional patent application, then the patent will receive the effective filing date established by the provisional patent application, with no substantive review of the provisional patent application.

So far, so good.

6.4 THE SPHERICAL PROVISIONAL PATENT APPLICATION

One of the most pervasive misunderstandings about provisional patent applications assumes that, because the formatting requirements are relaxed, so must be the substantive legal requirements. This is simply not the case:

> **While the formatting requirements of a provisional patent application are relaxed as compared to a nonprovisional patent application, the substantive (main) legal requirements are not.**

Two of the main substantive legal requirements that people tend to neglect when drafting a provisional patent application are the enablement and written description requirements. The confusion about these requirements likely derives from the fact that a provisional patent application is not required to have any claims. As we now know, the enablement and written description requirements are both measured with respect to the claims, which you recall are the terse statements at the end of the patent that set forth the boundary lines of the invention. In fact, without claims, you simply cannot tell whether the enablement and written description requirements have been satisfied.

Let's assume, for example, that you file a provisional patent application on an invention relating to a simple type of electronic game and then later claim your invention in the nonprovisional patent application within the scope of description of the various embodiments of the electronic game. It is likely you will be covered with respect to the enablement and written description requirements. If, however, you later try to stretch the claims to cover a quantum computer, it's unlikely that the enablement and written description requirements will be satisfied.

Unfortunately, you don't often hear about what happens when these requirements aren't met because very few provisional patent applications ever get scrutinized. This lack of scrutiny creates a false sense of security that you can file just about anything as a provisional patent application. And, in a sense, you can. The problems begin when you file the nonprovisional patent application and then down the road the invention actually has enough value to someone that he or she is motivated to scrutinize the provisional application.

The patent laws and subsequent court cases have made it abundantly clear that the specification of a provisional patent application must comply with the written description and enablement requirements of a nonprovisional application. Put differently, a provisional patent application has to support *at the time it is filed* any claims that are presented in the follow-on nonprovisional patent application. An inventor cannot simply add that support later and still enjoy the benefit of the provisional application filing date.

In some cases, technical papers are filed directly as provisional patent applications to secure effective filing dates. Academic institutions are particularly fond of this filing approach because professors are constantly generating technical publications as part of their jobs. The problem with this approach is that technical papers tend to describe *results* generated by experiments or experimental devices or discuss *why* things work, but not necessarily *how* to make and use an invention. Such applications have a high risk down the road of being considered nonenabling or of not satisfying the written description requirement.

The provisional patent application stake in the ground can range from a telephone pole to a toothpick depending on how well the provisional patent application supports the patent claims that ultimately issue as a patent, usually many years later. Many technology businesses that should know better file provisional patent applications with flimsy descriptions of an invention and then fill in the details later. An informed court is not going to honor the effective filing date of a provisional patent application if the claims that turn up years later in the corresponding patent are not supported in the provisional patent application.

To reiterate, no one is going to challenge a patent's effective filing date based on a provisional patent application unless the invention of the subsequent patent matters to a competitor. Then, the patent's effective filing date will be one of the first items scrutinized because that effective filing date in large measure defines what prior art can be considered in an attempt to invalidate the patent.

The question, then, naturally arises: What happens to an issued patent when its effective filing date based on a provisional patent application gets nullified?

6.5 INTERVENING PRIOR ART

As we discussed previously, the primary motivation for obtaining an early effective filing date is avoiding potentially problematic prior art that in most IP spaces is constantly popping up in the form of patents and publications.

Another motivation is avoiding self-inflicted damage due to the bars to patenting that can arise owing to a public disclosure of the invention prior to filing a patent application that covers the invention.

Let's take a look as some different filing scenarios to give you a better sense of why properly establishing an effective filing date matters. Figure 6.1 is a time line that shows a provisional patent application (PPA) filing on June 15, 2010 and a nonprovisional patent application (NPPA) filing on June 13, 2011, just shy of 1 year later. Also shown is a patent P issuing on December 20, 2011, from the NPPA. A related-art reference (REF) from someone other than the inventor with a publication date of June 30, 2010, is also shown.

The assumed effective filing date of patent P as established by the PPA is June 15, 2010. This effective filing date precludes the assertion of the REF against the NPPA as prior art during its examination because the publication date of the REF came **after** the filing of the PPA.

Figure 6.2 is similar to Figure 6.1, but now the effective filing date established by the PPA has been nullified because, for example, the PPA was deemed to not support any of the claims of P. In this case, the effective filing date of P becomes the filing date of the NPPA (i.e., June 13, 2011).

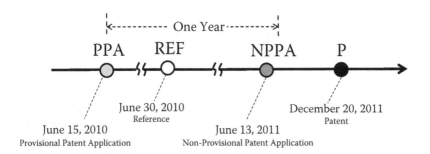

Figure 6.1 Time line illustrating the various dates in the example patent filing scenario when the patent remains "unmeasured."

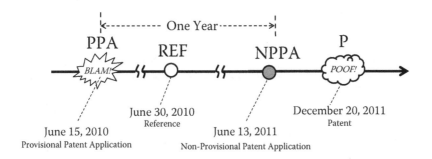

Figure 6.2 Time line similar to Figure 6.1, but with the provisional patent application effective filing date nullified, resulting in the destruction of the patent by the intervening prior art reference.

All of a sudden, it is as if the PPA never existed. The related-art reference (REF) can now be asserted against patent P as prior art to challenge its validity and perhaps blow up (that is to say, invalidate) the patent (or at least a problematic claim or claims). The REF is now called "intervening prior art" because it arose in the interval between the PPA and NPPA filing dates.

It is worth noting here that if the REF were actually by the inventor (or someone who obtained the information from the inventor), then the inventor would be entitled to the 1-year grace period from the date of the disclosure. In that case, REF is not prior art and the filing date of June 13, 2011 of the NPPA becomes the effective filing date of the patent P.

Figure 6.3 is similar to Figure 6.2 but also shows a date of July 20, 2009, denoted sale, that marks a public offer for sale date with regard to the claimed invention in question. Here we assume that the invention was disclosed in connection with the offer for sale by the inventor or the inventor's company. In this scenario, there is no potentially problematic REF. The effective filing date of June 10, 2010, of the PPA is within 1 year of the sale date, so problems with patentability arising from disclosure of the invention are avoided.

However, if the effective filing date of the PPA is nullified, then the new effective filing date is once again that of the NPPA—namely, June 13, 2011. This new effective filing date falls **beyond** the 1-year grace period for disclosures by the inventor (or someone who obtained the subject matter of the invention from the inventor), as measured from July 20, 2009. This means that the patent P (or perhaps just one or more claims thereof) can be declared invalid due to the on-sale bar.[1] Disclosures of the invention in other contexts can give rise to this same scenario.

What if there are no intervening prior art and no disclosure problems? Then the invalidation of the effective filing date of the PPA is moot. In this case, there was no real need for the earlier filing date in the first place. The whole point of a provisional patent application filing is to establish a placeholder for the earliest effective filing date because you can't really tell whether problematic prior art will show up or previously unreported invention disclosure activities will come to light.

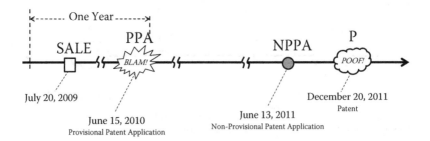

Figure 6.3 Time line similar to Figure 6.2, but with a sale/disclosure date for the invention destroying the patent when the effective filing date of the provisional patent application gets nullified.

6.6 THE PROVISIONAL-TO-NONPROVISIONAL CONVERSION PROBLEM

The safest provisional patent application is one that is drafted to resemble the follow-on nonprovisional patent application identically. This is because any differences between the two applications are going to be scrutinized to the letter if the resulting patent ever causes financial angst to someone else.

One problematic scenario that arises fairly frequently is when someone who is not an experienced patent attorney writes and files (or arranges to have filed on his or her behalf) a poorly drafted provisional patent application but then wants the follow-on nonprovisional patent application to be drafted by a patent attorney or patent agent. That attorney or agent is now going to have to strike a balance between describing the invention in the proper legal form and hewing to the text of the poorly drafted provisional patent application.

This balancing act becomes especially challenging when the drafter of the provisional application uses legally limiting phrases like "consisting of" rather than "including" or "comprising," fails to specify a variety of ranges for important parameters, neglects to describe the invention in terms of example embodiments, omits structures' key elements from the description and the drawings, does not consider alternative elements that might be used, does not draft sensible claims, or fails to address a litany of other seemingly minor but actually very important drafting considerations. Most often, the problem is the lack of details and specifics, based on the infamous IP "but I thought that" ("BITT") statement: "BITT including details would limit the scope of the invention and we want to have really broad claims."

As strange as "patentese" might seem to some, it is the accepted way of legally describing and claiming inventions and is the (still evolving) product of more than two centuries of patent practice. Contracts, wills, property deeds, and a host of other legal documents employ arcane legal language for the same reasons. While some think such language is a bunch of nonsense, the law is steeped in tradition and the concept of legal precedent, and making up your own legal terminology in place of that which is known to work is a recipe for disaster.

If the provisional patent application has no claims or if its claims were drafted by someone unfamiliar with the nuances of claims drafting, then the attorney drafting the nonprovisional patent application is going to have to add claims or try to revise the amateur claims. This can be highly problematic because the claims need to be supported by the description of the invention and the claim elements shown in the figures. Claims drafting is a back-and-forth process that synchronizes the language and phraseology in the claims with the description of the invention and the elements in the drawings. An improperly crafted written description in the provisional patent application is seriously going to hinder the drafting of claims later. As noted before, any discrepancies between the provisional and nonprovisional applications are going to create an opportunity for another party to attack the resulting patent if the patent is ever measured.

6.7 THE NEW-MATTER PROBLEM

A common scenario with provisional patent applications is when the provisional application omits a certain feature or element of the invention, but the follow-on nonprovisional patent application adds the feature or element. In patent jargon, this creates a "new matter" problem. The new matter added to the nonprovisional application is only entitled to the filing date on which the new matter was added—in this case, the filing date of the nonprovisional patent application.

Consider, for example, a provisional patent application for an apparatus that the inventor describes, in both the description and the drawings, as having a "mechanical switch." After the provisional patent application is filed, the inventor realizes that electronic and optical switches would work just as well and would be an easy way for competitors to design around the invention as described. The inventor's attorney, on behalf of the inventor, includes in the follow-on nonprovisional patent application reference to electronic and optical switches, and drafts claims that refer to, simply, a "switch."

The claims that refer to a generic switch with the intent to cover mechanical, optical, and electrical switches are unlikely to survive a challenge to the provisional application effective filing date. This is because the provisional patent application lacks support for all three types of switches. The electric and optical switch information constitutes "new matter." A dependent claim that limits the switch to a mechanical switch would be entitled to the effective filing date of the provisional patent application.

6.8 EFFECTIVE FILING DATE AND FIRST TO FILE

There is one last point about provisional applications and their effective filing dates that is worth discussing. As we mentioned before, most technology IP spaces are quite crowded with patents. Many different technology businesses are working with the same or similar technologies and are trying to patent the same or similar inventions. There are going to be many situations in which one technology business barely beats the other to the patent office and files first. Given that the United States now has a first-to-file system, the first inventor to file the patent application on the invention wins. It is game over for the filer who comes in second place.

But is it? If the invention at issue really matters (that is to say, if it represents substantial monetary value), then the technology business that came in second place, if it is smart, is going to take a very close look at the first-filed provisional patent application to see exactly what it discloses.

Consider now Figure 6.4, which is a time line similar to that in Figure 6.1, but now includes a competitor provisional patent application (CPPA) with a filing

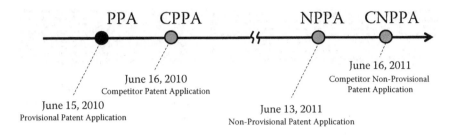

Figure 6.4 Time line similar to Figure 6.3, and showing, along with Figure 6.5, how a competitor's provisional patent application filed *after* the "first-filed" provisional patent application can become the first-filed provisional patent application.

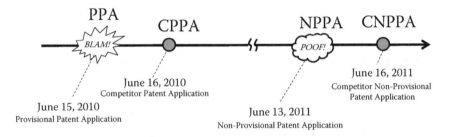

Figure 6.5 Time line showing how the regular patent application goes "poof!" once the effective filing date of the provisional patent application goes "blam!" Thus, the competitor patent application becomes the first-filed patent application.

date of June 16, 2010—just 1 day after the filing date of the PPA. Also, there is no related-art reference REF to worry about, and no patents have issued. The CPPA gives rise to a corresponding competitor nonprovisional patent application (CNPPA) filed on June 16, 2011, on the very last day of the 1-year filing deadline.

As you can imagine, with all the people rushing to the patent office to file their patent applications first, some are going to try to take a short cut. If that first-filed PPA falls short of meeting all the patent requirements, it is likely that sparks will fly over its effective filing date if the invention is worth something to someone else. The person who ends up filing "first," in other words, does not get the filing date etched in stone. If the invention ends up having real or perceived commercial value, then that effective filing date is sure to be tested.

If the stake in the ground representing the effective filing date of the PPA gets yanked out (say, by the owner of the CPPA) then the PPA first-to-file date of June 15, 2010, goes away, making the CPPA the first to be filed.

Figure 6.5 is the time line of Figure 6.4 in a later state of evolution and shows how the NPPA goes "poof!" as soon as the dubious PPA goes "Blam!"

It is worth reemphasizing here that what goes "poof" is one or more of the patent claims, depending on the nature of the problem with the PPA.

6.9 THE SCENARIO FROM CHAPTER 5 REVISITED

While we are considering filing time lines, it is worth taking a moment to revisit the scenario we mentioned in Chapter 5, where someone who is first to invent and also first to file can lose out to a second person to invent and to file. Let's take a look at the simple but profound time line of Figure 6.6.

In this scenario, on April 1, 2013, inventor A invents invention X and keeps it secret. Then B independently invents invention X and publicly discloses it—say, by publishing a technical paper on May 1, 2013. Then inventor A files a patent application on invention X on June 1, 2013. That patent application can be a provisional or nonprovisional application. The filing of A's application can be either without any knowledge of B's publication or with knowledge and in a panic to try to beat B to the patent office. Then, on July 1, 2013, B files a patent application on invention X. So to be clear, A is the first filer and B is the second filer. A is also the first to invent invention X.

Under the AIA laws that redefined what constitutes prior art, A's filing does **not** preclude B's later filing. In fact, A's filed patent application is not considered prior art to B's patent application. On the other hand, B's original disclosure on May 1, 2013, is prior art to the first-filer A's patent application. And since it discloses the same invention X, it precludes A from obtaining a patent on invention X.

Figure 6.7 is the time line of Figure 6.6 when B's disclosure finally catches up as prior art with A's patent application in the patent office and blows it up. At the same time, B's patent application remains intact because B's own disclosure is within the 1-year grace period of B's patent application filing. The result is that B's patent application prevails even though A was the first to file a patent application on the invention. This is why the US system is a "first to file" system only to within a spherical patent approximation.

Under the laws set forth in the new version of 35 USC §102 of the AIA, when an inventor or someone who obtained the subject matter directly or indirectly from the inventor (say, the technology business where the inventor is employed) discloses the invention, the disclosure inoculates the inventor patent application

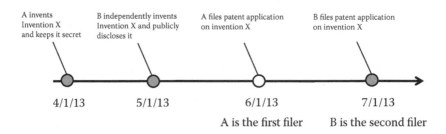

Figure 6.6 Time line of the scenario mentioned in Chapter 5 where the first inventor to invent and to file loses out to someone who invented second and filed second.

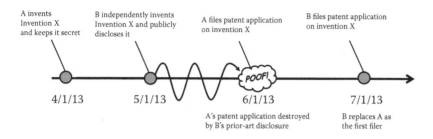

Figure 6.7 The time line of Figure 6.6 wherein B's disclosure serves as prior art that catches up with and blows up A's first-filed patent application.

filed later (if within a 1-year grace period from the first date of disclosure) from someone else's published art arising after the disclosure date. This is true even if the other person's published art was independently developed and even earlier developed.

All of this is to show that the time line of events associated with a given patent application or patent are important and subject to some fairly nuanced laws. The various time line dates and the contents of the various documents involved all will matter and be scrutinized if the invention turns out to matter to someone other than the patent or patent-application owner.

6.10 HOW BAD PROVISIONAL PATENT APPLICATIONS HAPPEN TO GOOD PEOPLE

Provisional patent applications are often used to facilitate a last-minute filing of a patent application in the face of an impending deadline. Often, this rush to file is precipitated by the sudden revelation that the invention is about to be disclosed at a trade show or to a customer or vendor. The fear that a competitor will file first also gets people rushing to the patent office. Sometimes, the rush comes from a completely artificial deadline generated by the craziness of a company's internal, high-entropy patenting system. Generally speaking, most (but not all) rush filings of provisional applications can be traced to patent system entropy in the technology business and a management-by-crisis mind-set surrounding IP matters.

The previously mentioned misconception about the informality of provisional patent applications tends to make people think they can cut corners. In an increasingly busy workplace, provisional patent applications can be rather seductive in this regard. They are a great way to postpone work for about a year. In fact, that is exactly what many technology businesses do. They don't check in 6 months out or even 8 months out or 10 months out from the provisional application effective filing date to initiate the preparation of the nonprovisional (and/or international) patent applications that would rely on the effective filing date of the provisional patent application. Often, this process is initiated just weeks prior to the 1-year deadline. So the manic rush to file the patent application picks up where it left off, with the provisional patent application, instead of being dealt with far enough in advance to avoid working in crisis mode all over again.

A low-entropy patenting system minimizes the number of rush provisional application filings and follow-on rush nonprovisional application filings. In most cases, the decision to file a follow-on nonprovisional patent application from a provisional patent application can be made well in advance of the 1-year deadline. It just takes having in place a best-practice IP system that makes good planning part of the process.

The education and training of technology workers can also help. Too often the patent application process is artificially plunged into crisis mode because a technology worker lacked the IP awareness to pass along, in a timely way, key information such as an upcoming date for a disclosure of the invention. A technology worker's IP knowledge needs to extend not only to the preceding information about provisional patent applications but also to the business-related triggers that would require action. To this end, technology workers who interface with customers and vendors or who attend trade shows need to understand how their activities can set legal time lines in motion and appreciate the importance of communicating pertinent information to the proper legal contacts.

6.11 IT'S NOT ALL BAD NEWS

Provisional patent applications are an important tool in the patent tool box. They just need to be understood and used correctly. When properly used, they can be indispensable when faced with a short time line, which always seem to happen, even in well-run organizations. They are also useful for establishing an early filing date while allowing time to assess the market for an invention before committing to the costs associated with a nonprovisional patent application as well as international patent applications. The point of this chapter is not that one should never use provisional patent applications. A provisional patent application can still meet all of the legal requirements while having informal formatting. The key is to understand that it needs to be constructed properly if it is to provide an effective filing date that can withstand a legal challenge.

NOTE

1. This scenario played out in an important court case relating to provisional patent applications, *New Railhead Manufacturing L.L.C. v. Vemeer Manufacturing Co.*, 298 F.3d 1290 (Fed. Cir. 2002).

CHAPTER 7

THE DOUBLE-EDGED SWORD OF INFRINGEABILITY AND VALIDITY

7.1 INFRINGEABILITY–VALIDITY BALANCE

One way to think about patent value is by how much the invention will contribute to the good of society. Another way to think about patent value is based on the discounted cash flow of royalties received or the costs avoided (that is to say: Should the money be spent on the patent or just put in the bank instead?).

A more legalistic and perhaps cynical way to think about the value of a patent is the one we touched on briefly in Chapter 4—namely, its likelihood of being infringed by someone of significance, such as another technology business, and in particular a competitor. We refer to this characteristic as *infringeability.* To a first approximation, infringeability depends on the breadth of the patent's broadest claim. Generally speaking and assuming a spherical patent, the broader a claim is, the greater the IP space that is owned and the greater the probability that others will infringe (trespass) upon it.

But (and again, assuming a spherical patent), broad claims are also more likely than narrow ones to be found invalid. A broad claim, for example, is more likely to encompass prior art. It also requires a higher level of enablement because the description of the invention has to enable the full scope of the broadest claim. So patent validity and patent infringeability form a double-edged sword. What you tend to gain with one you tend to give up with the other. And often, what you seek to gain from one comes back to haunt you in the other.

There is a balance, then, between a patent's infringeability and its validity. Or, more accurately, there is balance between a *claim's* infringeability and its validity, since patent infringement is assessed based on the claims, and it takes only one claim to find "patent infringement." Patent infringement is determined by the broadest claims in the patent, which are the "independent" claims, or the claims that stand alone and that do not refer to another claim. A set of claims usually also includes dependent claims that refer back to an independent claim and that by definition have a narrower scope than the independent claim on which it depends. In Chapter 9, we take a closer look at the distinction between independent and dependent claims and the role dependent claims play with respect to the independent claims.

For now, we take a closer look at how this infringeability/validity balance works in practice with respect to broadly crafted independent claims and explore practical issues pertaining to the fuzzy nature of patent claims, patent quality, and why the USPTO sometimes grants bogus patents.

7.2 THE INFRINGEABILITY–VALIDITY CURVES

Figure 7.1 depicts the infringeability–validity balance in the form of an infringeability/validity (I/V) graph. The I/V graph plots how likely a claim is to be infringed (i.e., infringeability) as the *probability of infringement P(Infringement)* (left-hand axis; dashed curve) and how likely a claim is to be found valid as the *probability of validity P(Validity)* (right-hand axis; solid curve). The curves for P(Infringement) and P(Validity) are plotted as a function of the number of meaningful limitations in a claim. The I/V graph features five zones. At the left end is zone 1, which represents a claim with no meaningful limitations. At the right end is zone 5, which represents a claim with an extremely large number of meaningful limitations.

To appreciate what the different zones in the I/V graph represent, it is helpful first to consider an actual invention—in this case, my very own laser toaster, which I hereby bequeath to the public. See if you can stand to read the following quasi-patentese description of my laser toaster invention. If you can endure it, you'll be better able to appreciate how the claims for the different zones relate to this description. Note also how the description sounds quite believable even though I confess to having no idea if the laser toaster would actually work.

Laser toaster detailed description

Figure 7.2 shows a perspective view of a laser toaster 10 that looks remarkably like a conventional toaster. Figure 7.3 is a close-up cross-sectional view that shows the internal workings of laser toaster 10. The laser toaster 10 has a conventional toaster housing 20 with a top portion 22 that includes slots 24. The slots 24 are sized to accommodate a slice of bread 40 to be toasted. A lever device 30 is used to adjust the position of bread 40 within

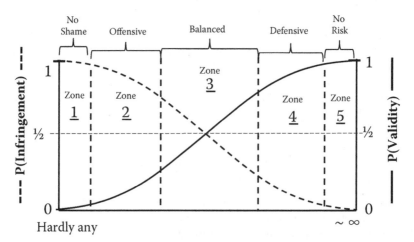

Number of meaningful claim limitations

Figure 7.1 Infringeability/validity (I/V) graph.

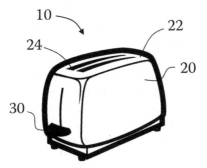

Figure 7.2 The laser toaster, which looks remarkably like a conventional toaster.

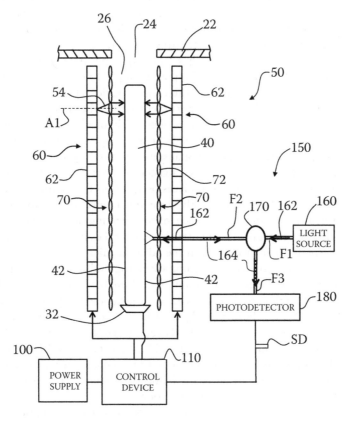

Figure 7.3 A close-up, cross-sectional view of one of the heating assemblies of the laser toaster.

a given slot 24 and is operably connected to a movable bread support member 32 that resides within slot 24.

The internal view of Figure 7.3 shows one of two heating assemblies 50 that are operably supported within an interior 26 of housing 20. The heating assembly 50 includes opposing and spaced-apart arrays 60 of infrared heating elements 62 in the form of solid-state heating devices, such as laser diodes or light-emitting diodes. Each heating element has an axis A1. Infrared heating elements 62 emit infrared light via a directional solid-state light-emitting process.

The heating assembly 50 includes opposing and spaced-apart micro-lens arrays 70, with each microlens array comprising microlenses 72. The microlens arrays 70 are respectively arranged between laser diode arrays 60 and are generally parallel thereto. The microlenses 72 are made of a material that transmits infrared radiation. An example material is germanium oxide. Each infrared heating element 62 is associated with a corresponding microlens 72, which lies along the corresponding heating element axis A1.

The bread 40 preferably has the form of a planar slice with substantially parallel opposite surfaces 42. The bread 40 is inserted into slot 24 so that it is supported by moveable bread support member 32 (and standoffs, not shown) and resides between the opposing microlens arrays 70. The bread surfaces 42 are preferably within 1 cm of the adjacent microlens arrays 70. This bread position is called the "toasting position."

The heating assembly 50 also has a power supply 100 electrically connected to a control device 110. The power supply 100 may be external— for example, an electrical outlet. The control device 110 is also operably connected to moveable bread support member 32. The control device 110 includes a timer (which can be electrical or mechanical) that causes moveable bread support member 32 to slide bread 40 up through slot 24 when the toasting process is completed or at some set threshold time.

Further in the operation of laser toaster 10, lever device 30 is pushed down to an "on" position, which places bread 40 in the toasting position. This causes control device 110 to allow electrical power to flow to infrared heating elements 62. In response, each infrared heating element 62 generates a diverging but directional beam of infrared light (i.e., heat) 54 that reaches the corresponding microlens 72 in the adjacent microlens array 70. Each microlens 72 takes the corresponding diverging and directional infrared light beam 54 and forms a corresponding collimated light beam, which is made incident upon bread surfaces 42 at a substantially right angle. Thus, microlenses 72 make the infrared light beams 54 even more directional.

In an example, control device 110 is configured to cause infrared heating elements 62 to emit pulses of infrared light 54 at a rate of 1 Hz or faster so that the bread surfaces 42 do not have time to cool very much between successive pulses. In another example, heating elements 62 operate in a nonpulsed (i.e., continuous) radiation mode.

In an example embodiment, laser toaster 10 includes an optical fiber toasting sensor 150 configured to detect the degree to which at least one of the bread surfaces 42 is toasted. The optical fiber toasting sensor 150 includes a light source 160 optically coupled to a circulator 170 via a first optical fiber section F1. A second optical fiber section F2 is also optically coupled at one end to the circulator 170 and is arranged so that its other end is in proximity to one of the bread surfaces 42. In an example, the end of second optical fiber section F2 that is closest to bread surface 42 is encased in a ceramic sheath (not shown) to shield the optical fiber from heat radiating from heating elements 62. A third optical fiber section F3 optically connects circulator 170 to a photodetector 180. In the operation of optical fiber toasting sensor 150, light 162 from light source 160 passes, in order, through first optical fiber section F1, circulator 170, and second optical fiber section F2, to illuminate a small portion of bread surface 42.

The light 164 reflected by bread surface 42 is then captured by second optical fiber section F2 and passes to circulator 170. The circulator 170 then routes light 164 to third optical fiber section F3 and to photodetector 180. Photodetector 180 then generates an electrical signal SD representative of the amount of reflected light 164 detected and provides this signal to control device 110. The amount of reflected light 164 is presumed to be inversely proportional to the degree of toasting. The control device 110 is configured (e.g., via a look-up table or other software embodied in a computer-readable medium) to assess the degree to which bread surface 42 is toasted based on the information in electrical signal SD.

So, setting aside the fact that my laser toaster is likely to cost about $20,000, be the size of a college dorm refrigerator, need a three-phase 50 W plug, and have a rather limited market, let's look at how we might claim this invention based on the different claim zones in Figure 7.1.

7.3 CLAIM ZONES

7.3.1 Zone 1

Zone 1 represents a claim so broad that it can't actually be distinguished from the prior art. However, its advantage is that it promises to capture the greatest number of infringers so that it has a high infringeability and thus "high value" by having strong licensing potential as well as strong protection for the invention by being able to keep so many others out of the market. I like to call a zone 1 claim a *no-shame claim*.

No-shame claims come in two main types: the *short-form no-shame claim*, which is remarkably terse and tends to read like haiku, and the *long-form no-shame claim*, which is the length of a reasonable claim but includes lots of meaningless limitations that tend to obfuscate the actual gist of the invention.

A short-form no-shame claim is relatively easy to spot and the indignation it evokes sets in quickly if you know the technology. Here's an example of how our laser toaster might be claimed as a short-form no-shame claim:

> *A toaster for toasting a slice of bread having opposite surfaces, comprising: a toaster housing having a slot sized to accommodate the slice of bread; opposing heating elements operably disposed relative to the slot and that radiate heat to toast the respective bread surfaces when power is applied to the heating elements.*

Now, you can't argue with the fact that this is what the laser toaster does. This would be a wonderful claim to own because it describes essentially any toaster that any toaster company could hope to build now and deep into the future. Were this claim valid, one could collect royalties from any company or end-user (i.e., infringer) that sold or used toasters. So from an infringement viewpoint (and thus licensing-potential viewpoint), life couldn't be better.

The problem of course is that this claim is woefully invalid. Why? Because it encompasses almost every toaster ever built. It's probably just a matter of time before archeologists discover that Cleopatra had such a toaster. The no-shame claim is far too general given the actual description of the invention. It is impossible to distinguish the laser toaster no-shame claim from a conventional (if not antediluvian) toaster. Yet, the description is not inaccurate; it is just ridiculously overbroad. It is an abstraction of what was actually invented. And yet, some patents have such claims.

Now consider a long-form no-shame claim for the laser toaster invention:

> *A toaster for toasting a slice of bread having a surface, comprising:*
>> *a toaster housing that defines an interior and that includes an upper surface having a slot open to the interior and sized to accommodate the slice of bread;*
>> *a moveable support member configured to movably support the slice of bread within the slot and being lockable in a depressed position;*
>> *at least one heating element that resides adjacent and spaced apart from the bread surface when the slice of bread is operably supported in the slot by the movable support member; and*
> *a control device configured to control an amount of power provided to the at least one heating element, thereby causing the at least one heating element to emit infrared radiation that is made incident upon the bread surface, causing the bread surface to heat over time to the point where the bread surface becomes toasted, wherein the control device mechanically engages with the movable support member and releases the movable support member from its locked depressed position to cause the slice of bread to at least partially extend from the slot at the upper surface of the toaster housing.*

Impressive, right? If you aren't paying attention, the long-form no-shame claim can start sounding like it's actually describing something new. While one part of your brain is telling you that this is nothing new, another part of your brain is saying, "With all this fancy language, there must be something here."

Eventually, you come to your senses and realize that the long-form no-shame claim is just as woefully invalid as the short-form no-shame claim is. While it lists many limitations, they are *absolutely meaningless* when it comes to distinguishing our laser toaster from a conventional toaster. All toasters have "at least one heating element" that when heated up emits—you guessed it—*infrared radiation,* which is a fancy way of saying "heat." The claim, by the way, does not identify the source of this heat as "electrical power" but simply as "power," which could be solar power. This means that the heating element could be a rock heated by the sun (again, like Cleopatra's toaster). The "moveable support member" that's been used for over a century or so to control the bread position and to push the toast out of the toaster is arguably superior to other methods of removing the toast— for example, the not-recommended method of removing the toast with a metal object when the toaster is plugged in.

The long-form no-shame claim is more insidious than its short-form counterpart because it resembles a reasonable claim and creates the illusion of validity. Long-form no-shame claims are camouflaged by superfluous patentese, so they have a greater chance of being issued by the USPTO via bureaucratic quantum tunneling, a phenomenon we discuss later. And like any claim in an issued patent, no-shame claims are entitled to the presumption of validity until the patent is challenged.

Not all no-shame claims are drafted with the express intent to offend. They can arise by accident, such as when those involved in the patent application make incorrect assumptions about the density of the prior art in the IP space. And not all patent owners would actually want to have no-shame claims in their patent. Many sensible patent owners would rather have valid claims of reasonable breadth than ultrabroad but invalid claims.

A good place to find no-shame claims is in published patent applications. Many attorneys, either on their own or on instructions from their client, start out with one or more no-shame claims, with the idea that the claims will get narrowed during the course of the examination. Of course, if they don't get narrowed, then the patent issues with the one or more no-shame claims.

IP-BITT: "BITT if we practice the prior art, we can't be an infringer"

Now at this point you might be tempted to think that infringing a no-shame claim is easily avoided simply by pointing to the massive amount of prior art and saying, "I'm just doing what everyone else has been doing for years." The "but I thought that" version of this sentiment that many patent attorneys hear is: "BITT if we practice the prior art, we can't be an infringer."

While eminently sensible and logical, this argument is not a defense for patent infringement. That is to say, it does not allow you to avoid being an infringer. That argument is used to *invalidate the patent claim being infringed*. In this way, a no-shame claim automatically puts an "infringer" on the defensive because if the infringer is in fact infringing the claim, even if he or she is practicing the prior art, the burden is on the infringer to overcome the presumption of validity to invalidate the claim and thereby avoid infringement. Infringement goes away only when the claim or claims being infringed go away, and that requires a finding of invalidity or unenforceability (the latter usually being associated with inequitable conduct by the patent owner or someone else connected to the patent).

This is why no-shame claims can be very problematic. When they are asserted by a patent owner bent on getting something from his or her patent, it can take a great deal of time and effort to invalidate the problematic claim.

7.3.2 Zone 2

A zone 2 claim still seeks to capture an unreasonable number of infringers but adds some meaningful limitations that reduce the claim's shame factor. A zone 2 claim is still claiming the invention aggressively and offensively. While a zone 2 claim is not anticipated outright by the prior art, it is hyperobvious in view of the prior art. I like to call a zone 2 claim an "offensive claim" because of the double entendre.

We will discuss later in Chapter 10 an example zone 2 claim—namely, claim 4 of US Patent No. 6,237,565, which is reproduced below, and which you can read now to see if you can get through the patentese to the gist of the invention:

> **4.** A vehicle control pedal apparatus (**12**) comprising:
>
> a support (**18**) adapted to be mounted to a vehicle structure (**20**);
>
> an adjustable pedal assembly (**22**) having a pedal arm (**14**) moveable in force and aft directions with respect to said support (**18**);
>
> a pivot (**24**) for pivotally supporting said adjustable pedal assembly (**22**) with respect to said support (**18**) and defining a pivot axis (**26**); and
>
> an electronic control (**28**) attached to said support (**18**) for controlling a vehicle system;
>
> said apparatus (**12**) characterized by said electronic control (**28**) being responsive to said pivot (**24**) for providing a signal (**32**) that corresponds to pedal arm position as said pedal arm (**14**) pivots about said pivot axis (**26**) between rest and applied positions wherein the position of said pivot (**24**) remains constant while said pedal arm (**14**) moves in fore and aft directions with respect to said pivot (**24**).

Here is the distilled version:

> A pedal that pivots, with an electronic sensor that measures a pedal arm position as the pedal pivots.

Claim 4 was invalidated by the Supreme Court for obviousness in *KSR Int'l Co. v. Teleflex, Inc.*, 550 US 398 (2007). While the actual claim was not invalidated for lack of novelty, it was considered about as close as one could get to lacking novelty while still falling into the "obvious" category, thereby making it a classic zone 2 claim. There is more about all this in Chapter 10.

A zone 2 claim for our laser toaster might read as follows (emphasis added):

> *A toaster for toasting a slice of bread having a surface, comprising:*
>> *a toaster housing that includes a slot that accommodates the slice of bread within a housing interior*
>> *a heating **array** that contains a plurality of directional heating elements that **each generates and directionally emits infrared radiation** when electrical power is provided thereto, the heating array being operably arranged within the housing interior and adjacent the bread surface when the bread resides in the slot*
>> *a sensor configured to detect an amount of toasting*
> *a control device operably connected to the sensor and configured to control an amount of electrical power provided to the heating array to control an amount of toasting*

Infringeability is still quite high and the validity of this claim remains highly suspect, but it is starting to sound a bit more reasonable. Now the heating elements are in an "array" and they "directionally emit" infrared radiation. These limitations make you stop and think whether they might distinguish the claim from a conventional heating coil used in a conventional toaster. These are the kinds of words lawyers will focus on when arguing to the patent office or to a court that the claim is novel and nonobvious.

Like a zone 1 claim, a zone 2 claim risks not being enabled to its full scope based on a lack of support in the patent's description. Our description of the laser toaster arguably does not support our zone 2 claim because we disclosed only a fiber-based sensor and not other types of nonfiber optical sensors, such as a noncontact laser thermometer.

Sidebar: Which Type of Claim Is Worse: Zone 1 or Zone 2?

It is worth taking a moment to consider the question of which claim is worse from the viewpoint of a potential infringer: A zone 1 claim or a zone 2 claim. You might be tempted to think that the zone 1 claim is worse because it is broader and therefore easier to infringe.

A zone 2 claim is actually much worse. A zone 1 claim can be more easily invalidated because it lacks novelty. Lack of novelty is established using a single prior art reference or a prior art device, which in our case would be a conventional toaster. It is much easier to show someone a single reference or device and then demonstrate how it includes all the elements in the zone 1 claim.

But a zone 2 claim by definition cannot be invalidated for lack of novelty. Its invalidation requires showing that it is *obvious* in view of the prior art. This is much harder to do because now two or more prior art items (e.g., publications, patents, devices, knowledge in the pertinent art, etc.) are need to be combined to make the case for obviousness. The obviousness analysis is more subjective than the novelty analysis. Even when the claim is barely novel, just being on the other side of the novelty threshold makes a huge difference in how hard it is to invalidate the claim. Good patent attorneys know how to formulate compelling arguments for nonobviousness as long as the facts stay outside the lack-of-novelty circle. Those nonobviousness arguments can keep a zone 2 claim in play for a long time. In the *KSR* case relating to obviousness that we discuss in Chapter 10, the zone 2 claim reproduced previously survived all the way to the Supreme Court.

7.3.3 Zone 3

A zone 3 claim seeks to balance what was actually invented with the known prior art. Its infringeability is substantially decreased relative to zones 1 and 2, but its probability of being found valid is substantially increased. An example zone 3 claim for our laser toaster might read as follows:

> *A toaster for toasting a slice of bread having a surface, comprising:*
> > *a toaster housing that includes a slot that accommodates the slice of bread within a housing interior;*
> > *first and second arrays of infrared radiation elements selected from the group of infrared radiation elements comprising light-emitting diodes or laser diodes, with the infrared radiation elements respectively directionally emitting infrared radiation when provided with electrical power;*

> an array of microlenses respectively configured and arranged to receive and direct the directionally emitted infrared radiation to the bread surface;
>> a fiber sensor configured to measure an amount of light reflected by the bread surface; and
> a control device operably coupled to the fiber sensor and configured to control the amount of power provided to the first and second arrays to control an amount of toasting.

While a zone 3 claim might try not to limit the scope of the invention unduly, it tries to respect the prior art boundaries by including limitations that seek distinguish it from the known prior art at least with respect to novelty. To this end, drafting a zone 3 claim generally requires conducting a prior art search to get an objective view of the IP space. Or, it is based on a good sense of the prior art that comes from staying up to date on developments (and patenting) in the given technology.

7.3.4 Zone 4

A zone 4 claim seeks to protect the invention in a conservative way by claiming exactly what the invention is without trying to abstract it to any substantial degree. Zone 4 claims are often used to cover inventions as they are used in commercial products, and they are sometimes called "product claims" or "defensive claims." A zone 4 claim has relatively low infringeability but now has a relatively high probability of being found valid.

A zone 4 claim for our laser toaster might read like the zone 3 claim but with some added limitations (in boldface):

> A toaster for toasting a slice of bread having first and second opposite surfaces, comprising:
>> a toaster housing that includes a slot that accommodates the slice of bread within a housing interior;
>> **first and second opposing arrays** of first and second infrared radiation elements selected from the group of infrared radiation elements comprising light-emitting diodes or laser diodes, with the first and second infrared radiation elements respectively directionally emitting first and second infrared radiation when provided with electrical power;
>> first and second arrays of **germanium-oxide** microlenses respectively configured and arranged to receive and collimate the first and second directionally emitted infrared radiation **to the first and second bread surfaces;**
>> **an optical fiber toasting sensor operably coupled to the control device and configured to illuminate one of the first and second bread surfaces and detect light reflected therefrom to determine the amount of toasting;** and
> a control device operably coupled to the optical fiber toasting sensor and configured to control the amount of power provided to the first and second infrared radiation elements to control an amount of toasting.

The added limitations make the claim easier to avoid (i.e., to design around), which reduces its infringeability.

7.3.5 Zone 5

A zone 5 claim has so many limitations that someone desperate to infringe the claim could scarcely do it even if his or her life depended on it. A zone 5 claim will capture only a pathologically committed infringer and gives the patent owner the peace of mind that the patent will be almost certainly found valid if measured. A zone 5 claim is thus a "no risk" claim. Its infringeability is close or equal to zero. It is the antithesis of the zone 1 no-shame claim.

Zone 5 claims sometimes result from an interaction with the USPTO wherein the patent examiner will allow certain claims only if they are combined. As a last resort, the patent owner agrees to combine the claims rather than forego patent protection altogether. Many technology businesses just want to own patents and don't actually care that much about what is actually in the claims.

Zone 5 claims typically exceed the *hand test,* which involves placing your hand lengthwise along the claim to see if a portion of the claim sticks out from the bottom. Given the microscopic type used in a US patent, it takes a rather long claim to pass the hand test. With all those limitations, it is going to be hard to capture any infringers but, generally speaking, you would have a hard time proving that such a claim is invalid.

A zone 5 claim for our laser toaster would include every limitation possible, raising its chances of being found valid (if measured) to essentially 100%. That said, the chances of anyone ever wanting to measure it are essentially zero.

Why, then, would anyone want a patent with zone 5 claims? While an individual patent with zone 5 claims might not be worth much, we observed in Chapter 4 that as part of a patent portfolio, these patents contribute to the overall patent pressure and have nonzero value in that context.

A related reason why a patent owner might pursue a zone 5 claim is to ensure that no one else can patent the same or similar invention. Recall, when the patent issues and settles into its IP space, it changes the size of the forbidden region where inventions cannot be patented. In such a case, the claims serve a role similar to a defensive publication.

7.4 FAIRNESS AND THE INFRINGEABILITY– VALIDITY BALANCE

Patent claims that represent a sensible balance between infringeability and validity are key to a fair patent system. The gross imbalance represented by zone 1 no-shame claims and offensive zone 2 claims is one of the main reasons people rail against what they perceive as the unfairness of the patenting system. Such claims are fundamentally unfair and can wreak havoc on those who try to conduct business in a reasonable and honest manner.

Patent claims that are incommensurate with what was actually invented inspire rightful indignation. Strip the invention of its veneer of patentese, and it may prove to be clearly nothing new or a trivial variation of what's known.

Nevertheless, it is important to acknowledge that the real problem here is the system that allows such claims and not necessarily the people seeking them. After all, it is unrealistic to expect people to exercise self-restraint when they have a chance to grab more than their fair share. Remember that, to many patentees, broader claims translate into greater patent value and thus the potential to earn more money. For many in the business world, the concepts of "money" and "fair share" aren't related. It is also worth pointing out here that patent attorneys are obligated to follow the instructions of their clients unless the instructions violate ethical rules. So if a client wants ultrabroad claims and the USPTO is willing to allow them, the attorney might have little say in the matter.

7.5 CLAIMS FOR PIONEERING INVENTIONS

Viewing patent claims as falling within zones 1 through 5 is an act of generalization that assumes a spherical patent and thus applies to most but not all patents and patenting situations. While most inventions are incremental improvements over the prior art, there are some inventions that are truly pioneering. A pioneering invention is a huge improvement over existing technology and often disruptive to the existing technology. Pioneering inventions often have very short claims that can look like zone 1 and zone 2 claims.

Example pioneering inventions include the telegraph invented by Samuel Morse (US patent no. 1,647), an improved sewing machine invented by Elias Howe, Jr. (US Patent No. 4,750), acetyl salicylic acid (aspirin) invented by Felix Hoffman (US Patent No. 644,077), and the solid-state transistor invented by John Bardeen and Walter H. Brattain (US Patent No. 2,524,035).

The following is a copy of the first claim of US Patent No. 2,524,035:

> What is claimed is:
>
> 50   1. A circuit element which comprises a block of semiconductive material of which the body is of one conductivity type and a thin surface layer is of the opposite conductivity type, an emitter electrode making contact with said layer, a col-
> 55 lector electrode making contact with said layer disposed to collect current spreading from said emitter electrode, and a base electrode making contact with the body of the block.

This is a relatively short and broad claim. It covers a solid-state transistor, and since this patent issued, a gazillion transistors have been made. This pioneering invention earned its right to be broad. This claim deserved to be valid and at the same time provide strong coverage for its owner during its lifetime.

So be careful not to confuse a zone 1 or zone 2 claim with a short and succinct claim that represents a truly pioneering invention. That said, most of the time you won't be confused because only a very small percentage of patents cover truly pioneering inventions.

7.6 BUREAUCRATIC QUANTUM TUNNELING

Under a classical patent mechanics viewpoint, the USPTO never issues an invalid patent. As shown in Figure 7.4(a), the various patent requirements collectively serve as a barrier to patent applications with dubious claims.

But, as you now know (if you didn't already), patents of dubious validity issue from the USPTO all the time. That a patent application should be able to traverse the requirements barrier without meeting every patent requirement is anathema and a source of great angst to those who ascribe to a classical patent mechanics view of the IP universe. For some, the patent cognitive friction presented by this situation is unbearable.

However, under a quantum patent mechanics view, the issuance of invalid patents is natural because, as shown in Figure 7.4(b), dubious patent applications can *tunnel through* the otherwise impenetrable patent requirements barrier according to a phenomenon I like to call **bureaucratic quantum tunneling.**

Bureaucratic quantum tunneling is akin to the quantum tunneling phenomenon that occurs in nature and that is exploited in certain devices, such as tunnel junctions and optical waveguide couplers. In nature, quantum tunneling occurs when the barrier presented to a quantum particle such as an electron or a photon is high and seemingly impenetrable but also sufficiently thin.

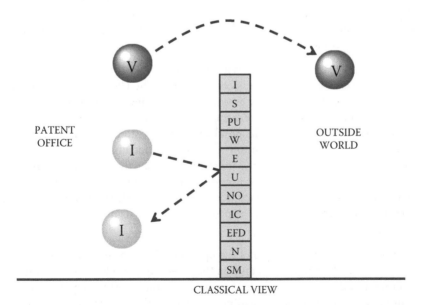

Figure 7.4 (a) Under a classical patent mechanics view, the patent requirements form an impenetrable barrier to any and all patent applications that do not meet the requirements.

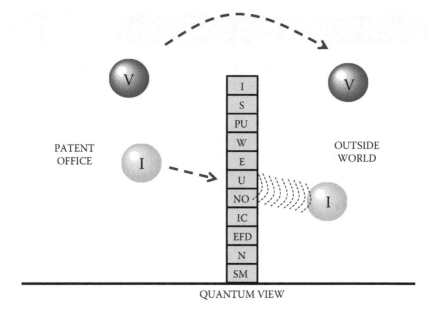

Figure 7.4 (b) Under a quantum patent mechanics view, bureaucratic quantum tunneling allows a patent application that does not meet the requirements to tunnel through the patent requirement barrier.

Along the same lines, the barrier formed by the patent requirements is certainly high and seemingly impenetrable. But the bureaucratic nature of the USPTO and the uncertainty inherent in the requirement filters also render the barrier thin and therefore potentially permeable under the right (or should we say *wrong*) circumstances.

Is this unfair? It is unfair like gravity is unfair. It is a function of the way the IP universe was created and has evolved and how the USPTO operates. Part of being a sophisticated technology business includes understanding that if you are in business long enough, you are eventually going to have to deal with a problematic patent. It may be that the problematic patent showed up at the company doorstep via bureaucratic quantum tunneling. Unfortunately, it is usually the smaller companies that suffer the most from bureaucratic quantum tunneling patents. There are many horror stories of small companies being bullied by other companies wielding their bureaucratic-quantum-tunneling patents and having to endure time-consuming and expensive litigation to deal with a charge of patent infringement.

The USPTO's recent institution of the option of the ***inter partes review*** via the AIA represents Congress's recognition that there were too few options available for quickly and affordably measuring a patent's validity. At least now there is an additional avenue available for dealing with potentially bogus patents before they get too far downstream. However, the inter partes review has its limitations and its critics, and as of this writing it is not yet clear how effective this option will be as compared to simply heading to federal court.

7.7 CLAIM DISTORTION

Students of science and engineering know that there are very few types of systems that can take an input signal and reproduce it with perfect fidelity. Usually, the output signal is distorted to some degree. Electrical and electronic systems, for example, exhibit a variety of distortions, including amplitude distortion, frequency distortion, phase distortion, and so on. Anyone who uses Skype knows all too well how a perfectly good electrical voice signal can be garbled and give rise to audio distortion. Optical distortion can be exploited to make the classic fun-house mirror. A more subtle kind of optical distortion known as diffraction arises when light travels through an aperture. Diffraction inherently limits the resolution of an optical system and can prevent the formation of a sharp image of the object.

The legal system is, well, a *system*. It receives information, processes that information, and outputs the processed information. It should come as no surprise that the legal system processes information with less than perfect fidelity. Moreover, the legal system is human based and therefore prone to emotion and subjectivity. But even on the planet Vulcan, where the legal system presumably has been stripped of all human emotion and is based entirely on logic, problems with distortion surely persist. These problems derive from limitations inherent in how information is processed.

The US patent system is part of the US legal system. A patent's claims represent the attempt to describe in words what the patent owner believes to be his or her legal property. Unlike real property such as real estate, which is tangible and can be surveyed complete with stakes in the ground, intellectual property is **intangible.** Trying to stake out the metes and bounds of intangible property is very difficult using words alone. As the American commonsense philosopher Josh Billings once said, "There's a great power in words, if you don't hitch too many of them together."

Unfortunately, drafting claims requires hitching words together in patentese. What's more, this word-hitching needs to create a clear boundary in IP space between the invention and everything else. It is nearly impossible to make such a boundary sufficiently sharp. Even those claims that essentially succeed at achieving a well-defined boundary will issue from the USPTO with a fuzzy boundary. The uncertainty in the examination process assures this distortion.

Figure 7.5 is a system diagram of how the USPTO and the courts process claims.

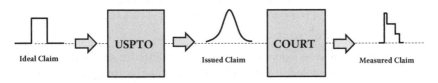

Figure 7.5 System diagram showing how the USPTO and the courts act as imperfect claim processors.

When a claim is drafted, it is intended to have a sharp boundary, like a sharp signal. However, when a claim "signal" is run through the USPTO, it issues with a smeared-out boundary because the uncertainty in the examination process gives rise to claim distortion. Only when a claim is measured by the legal system (i.e., by a court of law or the USPTO reexamination or inter partes review processes) is the boundary clarified. This clarification is the fundamental task of the court, which is charged with interpreting the claim and declaring where the boundaries lie relative to the dispute at hand. The court may not be much better at this clarification than you or I would be, but the difference is that the court's opinion is the one that matters.

7.8 CLAIM FUZZINESS

Figure 7.6 illustrates variations that can occur in the scope of a claim that otherwise appears to be well defined. The solid line represents the literal scope of the claim—that is, the claim as it is understood on its face. However, one or more terms in the claim may prove ambiguous. For example, consider a claim that uses the word "fluid." A fluid can mean either a liquid or a gas. But what if the patent application consistently refers to the fluid as a liquid? This usage supports a narrower definition of this term, which in turn would narrow the claim's scope beyond its plain meaning. This narrower scope is represented by the inner dashed line.

On the other hand, under the *doctrine of equivalents,* the scope of a claim can end up exceeding that of its literal meaning. The doctrine of equivalents is a legal gimmick that allows a court to stretch a claim beyond is literal meaning to prevent an apparent injustice to the patent owner that would go uncorrected were the law applied "with inexorable rigidity."[1] This claim stretching is represented in Figure 7.6 by the outer dotted line.

Often, people try to avoid infringement by making an insubstantial change to the claimed invention. Such changes typically include substituting a claimed element with an equivalent element—namely, one that operates in substantially the same way, performs substantially the same function, and achieves substantially the same result as the claimed element. Again, sometimes words fail to provide a sensible definition of the breadth of the property to which a patent owner should be entitled, and the court will balance the need for precise language to hold sway with the need to catch someone it thinks is trying to weasel the system.

Another legal maneuver is the flip side of the doctrine of equivalents. Called, not surprisingly, the *reverse doctrine of equivalents,* this maneuver can shrink the literal scope of a claim to ensure that inventions that differ substantially from the patented invention but that are somehow snagged by a literal reading of the claims do not get caught.

Consider, for example, a patent directed to an air filter for an air-conditioning system. The device includes a particle filter for filtering particles that travel through the air. The particles that make it through the particle filter hit a particle detector that counts the number of particles transmitted by the filter, thereby measuring the efficiency of the particle filter. One could imagine claim language worded in such a way that it might also cover an optical filter that filters particles of light in the form of photons, where some photons pass through the optical filter and are counted by a photodetector. Yet, the original air filter hardly anticipates the invention of the optical filter.

The reverse doctrine of equivalents is a somewhat odd legal gimmick because one would think that a claim that only accidently covered a dissimilar invention either would not be considered enabled or would fail to satisfy the written description requirement in regard to the dissimilar invention. Nevertheless, the reverse doctrine of equivalents is available for use on the behalf of accused infringers to shrink the literal scope of a claim in order to avoid infringement. A court case decided on the basis of the reverse doctrine of equivalents represents the classic situation wherein words fail to define the invention to such an extent that they unwittingly capture an infringer.

Figure 7.6 also illustrates how products can fall inside or outside the parameters of a given claim, depending on how the claim's scope is ultimately construed. Product 1 infringes claim 1 regardless of whether the literal scope of the claim is narrowed. Product 2 literally infringes claim 1 but avoids infringement when the claim's scope is narrowed. Product 3 avoids literal infringement but infringes claim 1 under the doctrine of equivalents when the claim's scope is stretched. Product 4 lies safely outside the widest possible margins of claim 1, so it does not infringe the claim either literally or under the doctrine of equivalents.

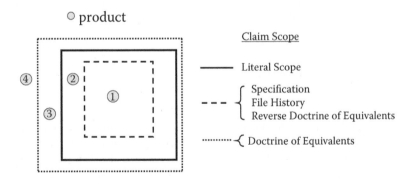

Figure 7.6 Schematic diagram illustrating the variation in the scope of a claim and showing how a product can fall within or outside the claim based on how the claim is construed.

As you might imagine, parties in a patent infringement lawsuit fight over how the claim should be read and, specifically, over how certain terms and phrases should be construed. And, indeed, how a court defines a claim's boundaries will have a huge impact on the outcome of the case. The part of a patent trial that construes (i.e., defines) the scope of the claims at issue is called a *Markman hearing* and is arguably the most important part of the patent litigation proceedings. In a Markman hearing, the judge hears arguments and then makes a ruling on certain words and phrases in a claim that are subject to different interpretations by the parties as to how they are to be construed. It is also called a ***claim construction hearing***. The Markman hearing usually happens early on in the case so that both sides can get an idea of where the cards may fall. Many a Markman hearing is soon followed by a settlement.

7.9 PATENT QUALITY

A concept related to patent validity and patent value is *patent quality*. For technology businesses, patents are like products and they need to have a quality standard. Patent quality speaks to how well the application was prepared, the zone or zones in which the different independent claims reside, how well the patent would hold up against a challenge by a competitor, and so on.

Generally speaking, the quality of a product is proportional to the effort put into making it. Most of us know the difference between a mass-produced item that is good enough to fulfill a utilitarian purpose at a sensible price and an expensive hand-crafted item that reflects a high level of craftsmanship.

As we noted in Chapter 3, patents all look pretty much the same because of uniformity of their printed format. However, anyone who has read enough patents will tell you that patent quality varies widely. At one end of the spectrum are low-quality patents that are mass produced in the manner of costume jewelry. Further along the quality spectrum are patents that have the quality of a small appliance, such as a toaster oven. Next come patents whose standard of quality matches that of a good car. At the high end of the quality spectrum are patents manufactured to aviation-industry standards.

Now here is the key point about patent quality:

> **A technology business gets to choose which quality standard its patents will meet.**

A technology business can choose to rely on the USPTO's quality standard and take its chances, or it can set its own internal quality standards. If a technology business chooses to rely on the USPTO's quality standards, then the quality of the resulting patents could lie anywhere along the quality spectrum. As we know by now, the USPTO's quality controls with respect to the numerous substantive requirements to obtain a patent are subject to substantial uncertainty.

A technology business can control the quality of its patents by setting its own quality standards. Choosing a high-quality standard requires maintaining a low-entropy patenting system whose internal processes are directed to tracking information about all the main patent requirements to ensure that they will be met should the patent ever be measured. It involves implementing IP best practices. Selecting a high-quality standard and then performing to that standard take a substantial commitment of time, money, and effort.

Choosing a low-quality standard and then performing to that standard is obviously much easier. The IP bazaar is full of folks that can help you with this endeavor. The low-quality route requires only low to moderate effort and is quite affordable. Remember, you can go to the IP bazaar and get a "patent pending" status today with a "provisional patent" for only $149.

Choosing an intermediate quality standard involves maintaining a sensible balance between time and effort on the one hand and expense on the other. The truism that you get what you pay for tends to hold when it comes to patents.

Not all technology businesses need to have patents of the highest quality. Many are in a position that allows them to mass produce their patents and not bother with aviation-industry-level quality controls. Their patents can have small-appliance quality or costume-jewelry quality because their ultimate value is vested not in the individual inventions but in the patent portfolio to which they belong. The portfolio can then be wielded as a patent-based WMD that exploits the benefit of the patent portfolio thermodynamics we discussed back in Chapter 4. So it can be a better choice for some technology businesses to spend less money per patent and not get hung up on quality so that they can boost the raw number of patents in the portfolio and increase the patent pressure of the portfolio balloon.

Remember, invalidating a patent of costume-jewelry quality still requires going to court or to the USPTO and usually takes substantial time, effort, and money. There is always the chance the district court will get the measurement wrong, which means an appeal to the CAFC for the losing side to get a better measurement. This in turn means more time, effort, and money spent on what might have seemed like a no-brainer way to make a bad and cheap patent disappear.

Some technology businesses that routinely seek low-quality patents are counting on the fact that an infringer would rather pay a nominal licensing fee than risk getting tied up in court. They are also counting on the probability that among a massive number of low-quality patents there will be some diamonds in the rough that are actually worth something. Remember, low-quality products that you buy on the cheap can be useful even if they break easily, as long as you have enough of them. Moreover, in the IP Era of the Information Age, having a huge stockpile of mass-produced patents is not only de rigueur, but is also useful for creating a scenario of mutually assured destruction that can make for a tenuous but peaceful coexistence between large patent-wielding companies.

NOTE

1. *Royal Typewriter Co. v. Remington Rand, Inc.,* 168 F.2d 691 (2nd.Cir. 1948), cert. denied, 335 U.S. 825 (1948).

CHAPTER 8

LOST IN IP SPACE

8.1 THE EXPONENTIAL GROWTH OF PATENTING

Data from the USPTO on the rate of patent application filings and the issuance of patents confirms what everyone in the patent business knows: The pace of patent application filing and patent issuance is on the rise and has been for an extended time. Figure 8.1 plots the number of patent applications filed and the number of patents issued per calendar year with the USPTO since 1940.

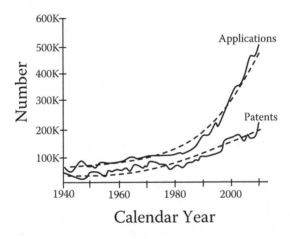

Figure 8.1 USPTO data of filed patent applications and issued patents since 1940.

Some believe that this increased pace is abnormal. They cite it as evidence of a "patent race" or "patent explosion" that reflects the sudden realization on the part of technology businesses that patents can have great value. While there is some truth to this, there is also a more fundamental dynamic at play that has to do with the nature of innovation. In a technologically driven society, inventions beget other inventions, which give rise to new and improved technologies that also expand and develop. As a result, the number of inventions available for patenting grows exponentially over time.

To understand how this exponential growth can naturally occur, let's consider a familiar invention: the automobile. The invention of the automobile in the 1880s was the big bang that opened up automobile IP space. As shown in Figure 8.2, the automobile gave rise to a first wave of automobile improvement inventions, such as tires, headlights, windshields, shock absorbers, engines, and so on. Each of these improvement inventions in turn gave rise to its own set of improvement inventions. For example, the invention of the windshield gave rise to a secondary

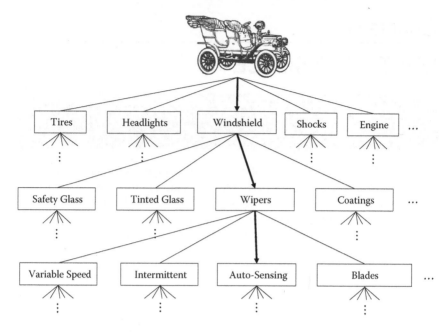

Figure 8.2 The automobile IP space illustrating the exponential growth of automobile-related inventions.

wave of windshield improvement inventions, such as safety glass, tinted glass, windshield wipers, anti-raindrop coatings, and so on.

Now consider just the invention of the windshield wiper. This invention defined its own IP subspace and gave rise to a wave of windshield-wiper improvement inventions, such as the variable-speed wiper, the intermittent wiper, the auto-sensing wiper, and heavy-duty wiper blades. There are companies whose sole business is to make and sell wiper blades. In fact, one of the most famous patent lawsuits in US history involved the invention of the intermittent wiper blade by Robert Kearns and his battle against Ford Motor Company. Kearns's legal battle over his invention was the subject of the 2008 movie *Flash of Genius*.

The same kind of ramifying trail can be followed with tires, headlights, shock absorbers, and almost every other part of the automobile. The process continues to this day for automobiles and every other technology worth improving. Each improvement invention opens up its own IP subspace that fills with improvement inventions that in turn open up other micro-IP spaces, then nano-IP spaces, and so on, seemingly ad infinitum.

The exponential growth of the number of patents being issued by the patent office is a fact of life in the present IP Era of the Information Age. It is an inherent characteristic of the IP space of a technology that has enjoyed commercial success. Consider how the invention of the modern computer was the big bang that created the computer IP space and all of its attendant technologies—semiconductor chip technologies, semiconductor equipment technologies, measurement technologies, display technologies, keyboard technologies, software technologies, telecommunications technologies, the Internet, and so on. Then think about how each of these technologies has given rise to its corresponding subtechnologies, about all the new companies that formed to fill the need for products in the various technology spaces, and about how each of these new companies needs to be patenting in order to maintain its hold on its own piece of IP space.

8.2 THE FRACTAL NATURE OF INNOVATION

The flow of improvement inventions toward increasingly finer and finer scales increases the density of an IP space. This densification process reveals the fractal nature of innovation. Fractals, so named by Benoit Mandelbrot in 1975, are amazing mathematical entities. Certain types of fractal patterns are "self-similar"— meaning that the pattern is formed from a basic shape that repeats on increasingly smaller scales.

We can zoom into or out of any part of a self-similar fractal pattern and see the same image. Fractal patterns are found almost everywhere in nature, from the structure of mountains, to the flow of water, to the form of animal markings. Once you know about fractals, it is hard to look around without seeing something with a fractal pattern.

Let's consider from a fractal viewpoint how an invention opens up an IP space and then how that space gets filled. To do so, we'll make use of the fractal pattern called an *Apollonian gasket,* which is illustrated in Figures 8.3(a) through 8.3(f). In Figure 8.3(a), a new invention, such as the automobile, gives rise to a new IP space. This new IP space is represented by the empty Apollonian gasket. In Figure 8.3(b), we see that the core patent on the invention occupies a major

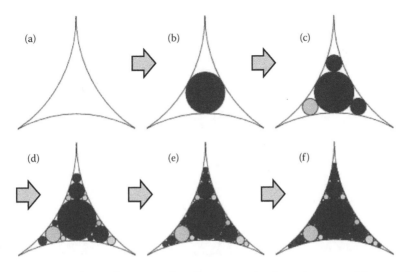

Figures 8.3 (a–f) Apollonian gasket illustrating the fractal nature of how an IP space fills with improvement patents. (Courtesy of Wolfram Research, Inc.)

portion of that space. Then, as shown in Figure 8.3(c), the first wave of improvement inventions comes along and their patents occupy the available portions of the IP space next to the main invention. As shown in Figure 8.3(d), a second wave of improvement inventions builds off the first wave and results in patents that fill in the gaps next to the first wave of improvement patents. This process continues, and the IP space grows increasingly dense, as shown in Figures 8.3(e) and 8.3(f). Pretty quickly the IP space becomes packed with subsequent waves of improvement inventions and their corresponding patents.

The amazing thing about fractal geometry is that no matter how dense the pattern appears, there is always some unoccupied space. Even in the densely packed IP space of Figure 8.3(f), the interstices have the same shape as the original IP space. So, in theory at least, an IP space can never entirely fill up, which means there is always room for innovation. From a practical viewpoint, however, an IP space can become so densely packed that whatever minuscule spaces are left in the interstices are commercially insignificant.[1]

In addition to zooming in on an IP space, you can also zoom out to get a picture of the larger IP universe and your place within it. Figure 8.4 is a larger Apollonian gasket packed with the smaller Apollonian gaskets of Figures 8.3(a) through 8.3(f).

If the IP space illustrated in Figures 8.3(a) through 8.3(f) represents the windshield wiper IP space, then the IP space of Figure 8.4 is the larger automobile IP universe. It is easy to think that your particular IP space is all there is and that you are at the center of it. It is far more likely that your IP space is just a subspace of some larger IP space, which in turn is just part of an IP multiverse.

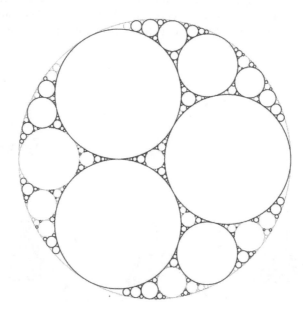

Figure 8.4 Larger view of an Apollonian gasket, illustrating how an IP space can be the subspace of a larger IP space. (Courtesy of Dr. Robert Lawson Brown.)

8.3 PATENTING IN DENSE IP SPACES

The fractal nature of innovation explains why innovation can occur even in very densely packed IP spaces. To the extent that improvements to a technology are commercially valuable, they can be generated on an ongoing basis at finer and finer scales as the improvements get more tightly focused.

Let's consider for a moment the rather densely packed IP space for light-emitting-diode (LED) technology. LED technology is a huge business, and these little light sources are being packaged for use in everything from airplanes to doorknobs. LEDs are rapidly replacing incandescent and fluorescent light bulbs. Twenty years from now, incandescent lights will seem as archaic as gas lanterns.

The LED IP space is older than most people think. LEDs are based on a property called electroluminescence, which was first observed in an electronic device—a Schottky diode formed from SiC (carborundum)—back in 1907.[2] The nascent LED IP space was filled with what is now core LED technology, such as the different types of materials and the different semiconductor configurations used to make LEDs a viable light source.

The modern LED IP space has, over time, fractally evolved and expanded into a huge number of IP microspaces directed, for example, to different ways of packaging LEDs to form light sources and to relatively small variations in the semiconductor structure that improve the LED's light output, wavelength stability, spectral output, efficiency, and so on.

For example, one recently formed LED micro-IP space contains a number of LED patents directed to the seemingly minor improvement of adding surface roughness to one of the semiconductor layers so that the structure scatters more light, thereby increasing the amount of LED light emission.[3] This seemingly small LED IP microspace will no doubt keep filling up and dividing into even smaller LED IP nanospaces as others invent different types of surface structures and unique methods of efficiently forming these structures.

As long as LED technology continues to garner substantial commercial interest, the LED IP space will continue to expand and divide into finer and finer scales of improvement inventions.

8.4 THE CLASSICAL IDEAL IP SPACE

Figure 8.5 shows a cartoon view of a portion of an IP space filled with some spherical patents that represent one form of prior art for any new invention that wants to enter the space as a patent. The size of each spherical patent can be thought of as representing the scope of the patent's claims. Also shown in the IP space are nonpatent prior-art blobs that represent publicly available information, such as technical information in journal articles, expired patents, existing products, and general knowledge in the art. We focus our attention on the two spherical patents that are separated by a distance D. These patents, while covering

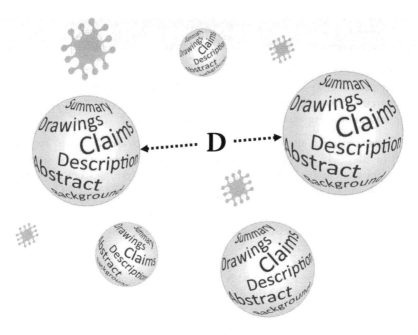

Figure 8.5 Cartoon view of an IP space that contains spherical patents.

very similar subject matter and therefore residing near each other in IP space, nevertheless have different claims and occupy different parts of the IP space.

If, to be patentable, an invention had only to be novel, then the spherical patents in the IP space could just touch each other and the prior art, as in the Apollonian gasket example of Figures 8.3(a) through 8.3(f). The distance D would be zero. As long as the claims were not identical, the spherical patents could be said to cover different inventions and thus could be considered novel with respect to each other and to the rest of the prior art in the IP space.

But in addition to the novelty requirement for patentability, there is the requirement that inventions be ***nonobvious.*** This means that one cannot patent an invention that has the absolute minimum amount of novelty relative to what has already been invented or is already known in the art. There must be some nonzero distance D between the new invention and the closest prior art that represents an additional level of inventiveness beyond sheer "newness."

The nonobviousness requirement therefore acts as a kind of repulsive force that arises from the mere presence of prior art in the IP space. When a new patent tries to enter the IP space, the prior-art force field prevents (or at least tries to prevent) the patent from residing too close to the existing prior art. It is a bit like trying to place a new, positively charged object within a preexisting arrangement of other positively charged and fixed objects. If you try to place the new object at a location in the electric field where the forces are unbalanced, then it will be pushed into an equilibrium location. In a similar manner, the repulsive force of all the existing prior art should, in theory, push a newly entering patent into an equilibrium position far enough away from the prior art for the new patent to be considered valid.

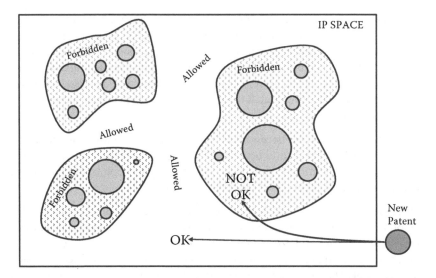

Figure 8.6 IP space diagram illustrating an ideal IP space where the prior art defines allowed and forbidden regions where patents can and cannot reside.

To be sure, it is perfectly fine to place a new patent at such a distance from the prior art that the prior-art force field is weak or nonexistent. Such remote placement might reflect a conservative patenting strategy that seeks to ensure patent validity. Perhaps the patent is directed to specific products that can be claimed narrowly.

This is to say that there is more than one position where the patent can reside and still be considered valid relative to the prior art, as illustrated in the simple two-dimensional IP space of Figure 8.6, where each circle in the space represents the scope of a prior-art reference (e.g., patent, publication, existing product, knowledge in the art, etc.). An invention that falls within one of the circles would lack novelty.

The prior art in the IP space gives rise to regions where a new patent cannot reside because it would be considered obvious. These obviousness-based *forbidden regions* in IP space are necessary for several important reasons. First, people must be prevented from taking advantage of the fact that claim language is imperfect and employing weasel words to patent the most trivial so-called "improvements" over existing inventions.

Second, others must be allowed the chance to design around patent claims by working in a forbidden region where inventions are considered obvious and therefore unpatentable. Designing around a patent requires that some breathing space exist around patents in the IP space so that obvious and noninfringing variants of a patented invention are available to the public. Said differently, one is allowed to use inventions in the forbidden regions because inventions that lie within are not subject to patent claims and are therefore available to the public.

The forbidden regions of IP space prevent the cost of products based on a given technology from skyrocketing from patent licensing fees. If every minor variant of the technology can receive the honor of a patent, the end products would be extremely expensive because patent royalties would weigh them down. Fairness dictates that the only technology businesses entitled to patent royalties are those that truly invented something worthy of a valid patent. People understandably get annoyed with the patenting system when they see the USPTO issue patents in a forbidden region and then watch those patent owners stick their hands out for a royalty payment.

The forbidden regions are necessary if innovation is to continue to the limits of commercial usefulness (and beyond, for those who patent without regard to commercial considerations) while ensuring that any tax one has to pay in the form of royalties to patent owners is reasonable and fair.

In a classical ideal IP space, all patents reside in allowed regions, so all are valid. The job of the USPTO is to make sure that each patent ends up in an allowed region in the IP space and not in a forbidden region. But as we now know, there is no such thing as a spherical patent, and neither is there a classical ideal IP space. The uncertainty that permeates the patenting process limits the USPTO's ability to place every single patent that it issues into an allowed region of the relevant part of the IP space. And that is when all the fun begins.

8.5 IP SPACE UNCERTAINTY

People who don't work in the patent field tend to underestimate just how densely most technology spaces are packed with patents. There is nothing inherently wrong with a densely packed IP space, as long as there are no patents in the forbidden regions.

Unfortunately, life in the IP universe is full of uncertainty. Most IP spaces have some patents in the forbidden regions. The USPTO unwittingly counters the naturally repulsive obviousness force field formed by the prior art when it jams a patent into forbidden region of an IP space. This upsets the ideal order of the IP universe—or, as Obi-Wan Kenobi might say, it creates a disturbance in the force.

Why does the USPTO do this? It comes down to resources. The USPTO just doesn't have the necessary resources and time to map out the prior-art force field for every IP space to the resolution needed to ensure accurate and proper placement of newly issued patents in the IP space. In many instances, it has a fairly low-resolution picture of the IP space compared to what one could obtain in theory with all the available resources and all the time and money in the world.

It is important to appreciate how hard it is really to understand and map an IP space to a detailed level. Patents can be hard to read and understand. Their boundaries are fuzzy. Also, some prior-art references are simply more accessible and more easily searched for than others are. For example, US patents are quite easy for the USPTO to find since it issues them and still has the piles of paper copies, as well as the PDF copies stored in its local database. It is no surprise that US patents are the documents the USPTO cites most often in office actions.

However, there are also many highly informative Japanese technical journals that publish on a wide range of subjects. There are also plenty of German patents on highly technical inventions. In fact, lots of countries publish patent applications, issue patents, and generate large numbers of technical publications. But such patents and technical publications are not always cited because they are simply not as easy to find when searched for using reasonable and affordable search methods.

The reason that the USPTO cannot form a clear picture of the IP space for every patent application is analogous to the reason that bank security cameras can never seem to take a clear picture of a bank robber. There exists optical technology that allows one to read the date on a newspaper from outer space, but this technology must be too expensive to implement in security cameras. Apparently, it is more cost effective to install low- to medium-resolution cameras where robberies might occur and hope the robber has a prominent tattoo or a bad disguise.

Likewise, it is prohibitively expensive to put together a comprehensive government-based prior-art searching system that has high resolution. Low to medium resolution is affordable and reveals the most conspicuous prior art. That key prior-art document sitting in a library in Norway is probably going to stay hidden until someone cares enough to conduct a scorched-earth (read: expensive) prior-art search to hunt it down.[4]

Anyone who has tried to conduct an in-depth prior art search knows how hard it can be. Vagaries in terminology alone can, in some technologies, hide enough prior art to fill a pickup truck. In addition, the term "prior art" encompasses articles, patents, technical publications, brochures, website information, and generally anything ever written in any country on the planet, not to mention knowledge in the art that can sometimes be impossible to find in a written document. Furthermore, IP space is dynamic, not static. New prior art enters all the time. So your measurement of the prior-art force field today might be different next week when a new piece of prior art (say, in the form of a newly published US patent with a filing date earlier than yours) enters the IP space and changes the force field.

A prior-art search is also highly subject to the law of diminishing returns, so the searching process tends to be asymptotic. Finding about half of the prior art that exists in the space is usually fairly easy to accomplish through straightforward searches of publicly available databases such as the USPTO website, Google Patents, Espace.net, and the like. Getting what your gut tells you is about 80% of the prior art can be difficult to downright strenuous. Getting 100% of the IP space mapped in hopes of removing all uncertainty about what prior art exists is next to impossible.

This state of affairs leads us to a corollary of the patent uncertainty principle that we'll call the "prior-art uncertainty principle":

> **It is extremely difficult if not impossible to know the prior art of an IP space with 100% certainty.**

The good news here is that as long as you know an IP space better than anyone else, you are usually ahead of the game.

8.6 IP BLACK HOLES

When almost every possible improvement invention in a technology gets patented, the number of patents that reside in the forbidden regions of the IP space can become extreme. This makes one wonder whether there is some fundamental limit to how densely packed with patents an IP space can be.

The short answer is yes, there is a fundamental limit, and it is set by the obviousness standard. At some point, a given IP space can't accept any more new patents because all the patentable space represented by the "allowed" region in Figure 8.6 has been taken up. All that is left is one big forbidden region. Since obviousness prevents patents from being packed right up against other patents and the prior art, even patents that the USPTO jams into the forbidden regions usually have some distance between them.

However, in some cases, the IP space can become so dense that not even a photon of innovative thought can escape or pass through. At this point, there is no longer a discernible forbidden region in the IP space. The patents are crushed up against one another at finer and finer scales and form what from a distance looks like one dark megapatent of infinite density.

The USPTO has allowed an IP black hole to form in IP space (Figure 8.7).

Figure 8.7 An IP black hole sucks innovation and money out of the IP space in which it resides.

In a normal if not ideal IP space, unpatented and unpatentable innovations reside in the forbidden regions and are available to the public despite residing next to some patents that also ended up in the forbidden regions. However, in the case of an IP black hole, every single possible improvement has been patented, including all those in the forbidden regions that should be publicly available because their obviousness made them unpatentable. But innovation cannot move through an IP black hole. Like a regular black hole that sucks in matter that gets too close and crushes it by gravitation, an IP black hole sucks in any innovation that gets too close and crushes it by litigation (or the fear of it).

In a regular black hole, time stops. In an IP black hole, innovation stops. Steven Hawking discovered that black holes can radiate. IP black holes can also radiate— they radiate angst,[5] and anyone who gets too close to the IP black hole starts to feel very nervous about patent infringement and being sued.

Products that reside in an IP space where an IP black hole exists can experience a huge IP tax owing to all the overlapping claims of the patents therein. The smartphone IP space is a classic example. There are so many patents crowded into the smartphone IP space (and its microspaces) that there is no way anyone new could enter this market without a disruptive technology that opens a new IP space. The way patents are packed into these kinds of IP spaces rivals the way the ancient Inca stone masons built their temples—without room for even a piece of paper to be slipped between the stones. The dizzying array of law suits associated with the mobile phone and smartphone IP spaces (as discussed briefly in Chapter 2) shows what can happen when there is a superdensely packed IP space in a wildly valuable technology.

Beyond the added tax, all the patent owners in such an IP space can end up either trying to sue each other or threatening to sue each other if they don't get their share of the pie. A substantial portion of the cost of a smartphone (by common estimation, 25%) goes toward paying patent royalties on the technologies embedded therein. So an IP black hole also tends to suck money out of the pockets of anyone whose technology gets too close.

Sidebar: Possible IP Space Black Hole

IP space astrophysicists believe they have spotted an IP black hole in the region of IP space that contains polishing-pad technology used in chemical mechanical planarization (CMP). CMP is used in the fabrication of semiconductor integrated-circuit (IC) chips. In brief, the IC chip fabrication process typically involves forming various material layers and structures over previously formed layers and structures deposited on a silicon wafer. However, the underlying features cause the top surface of the layered structure to be uneven, which makes performing photolithographic processes on this top surface problematic.

It is thus necessary to create a flat-top surface. The CMP process, which involves pressing the uneven top surface of the layered structure against a rotating polishing pad in the presence of abrasive slurry, is the conventional way to achieve such a surface. Of course, one needs to know when to stop polishing (i.e., the polishing end point) so that the layered structure beneath the top surface does not get completely eroded away.

It appears that the inventors on US Patent No. 5,893,796 were the first (or at least among the first) to come up with the idea of putting a transparent window in the polishing pad so that a laser could be directed through the window, reflect off the wafer layer being polished, and then reflect back through the window to the detector. One might think that once the '796 patent issued, anyone who tried to patent a polishing pad with a window would be out of luck.

Not so. Since the '796 patent issued, all kinds of patents with claims to CMP polishing pads with windows have issued. One type of polishing pad window has a recess. Another type has reduced stress. Another has high optical transmission, and yet another has an antiscattering property. One patent that issued long after the '796 patent has what appears to be a long-form no-shame claim that surely looks like it is simply claiming a window in a polishing pad. Another appears to have a short-type no-shame claim to—you guessed it—a window in a polishing pad. Some patents in this space claim polishing pads with windows of select materials. Every company in the CMP business seems to have squeezed into the minute interstices of the CMP-pad IP space its own version of the CMP polishing pad with a window. And patent applications on ever finer variants of the CMP pad–window combination are still being filed to this day, as indicated by recently published patent applications.

At the very least, the IP space for CMP polishing pads with windows illustrates just how finely parsed an IP space can become. At the very worst, it charts the formation of an IP space black hole where every imaginable variant of a window in a CMP polishing pad gets patented instead of being rejected as an obvious variant of the original invention.

8.7 WHAT ABOUT US?

Densely packed IP spaces can lead some companies to become unwilling *not* to file patents in the technology, even on inventions they believe to be utterly trivial. "After all," they say, "our competitors are filing patent applications on their 'obvious' inventions and getting them issued as patents. What about us? Why shouldn't we try to squeeze ours in as well so that at least we have something to negotiate with in the event we end up being sued?"

They have a good point. Unfortunately, this situation can result in a race to the bottom and cause IP space densification that leads ultimately to the formation of an IP black hole. This is not a good thing. But tell that to the company that doesn't have the patents. If your competitor is filling up the IP space with patents to your detriment, the most obvious countermeasure is to start seeking your own patents.

8.8 PATENT WAVES AND TECHNOLOGY WAVES

One IP "but I thought that" comment I hear far too often is "BITT the invention must be patentable because I've never seen it." The misguided presumption of this particular IP-BITT is that all patented inventions are embodied in existing products for all to see. After all, no business would patent something that was not going to be incorporated imminently into a real product, would they?

You bet they would. In fact, that is the norm. Most inventions are patented far in advance of when they might appear in a product. This is true even though a patent application can be filed up to one year after the invention is sold as a product. In the main, most patented inventions are speculative and never make it into a commercial product. In this sense, patents are a great way to see what people think might be valuable technology in the future. There are a lot of dead IP spaces filled with patents on inventions that turned out to be nonstarters as products.

For any serious technology, waves of patents precede the products based on the patented inventions, as illustrated in Figure 8.8(a).

These patent waves are like the pressure and shear waves associated with a seismic event. The pressure waves travel faster than the shear waves, so they arrive before the shear waves. The IP space of a technology will be fairly dense with patents well before the technology starts to get off the ground. The first or primary patent waves to hit belong to the big, low-frequency, broad-core patents, followed by the intermediate-frequency improvement patents. Then the IP waves start to decrease in size but increase in focus and frequency as the secondary improvement patents focus in on the commercial implementation inventions that make the technology commercially viable. Then the product arrives, well after the patent party started and the patents were established in the IP space. Of course, in some cases, the product never arrives owing to a change in plans, the lack of a market, and so on.

So here is a key concept about inventions that are apparently not embodied in any products:

> **The fact that an invention is not embodied in a product does not mean that there isn't one—if not dozens—of patents related to that invention.**

Take quantum cryptography as an example. Quantum cryptography offers the promise of perfectly secure encryption using single-photon light pulses and

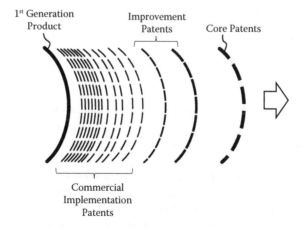

1st Generation Product   Improvement Patents   Core Patents

Commercial Implementation Patents

Figure 8.8 (a) Patent waves of core, improvement, and commercial implementation IP precede the first generation product.

the principles of quantum mechanics to ensure no one is eavesdropping on the communication link. Serious commercialization of quantum cryptography has yet to be realized, with only a couple of small companies offering a commercial product. Yet, the quantum cryptography IP space is already incredibly densely packed with patents of all kinds. And the IP space was crowded even before the first commercial product was offered. Lots of folks thought quantum cryptography was going to be a hot market in the 1990s and 2000s because the technology enables completely secure encryption. But the market didn't materialize, and now the quantum cryptography IP space sits all tiled up with patents in a state of near suspended animation as the arrival of the market is awaited. There are plenty of other markets and IP spaces just like this.[6]

It is also worth noting that, in many cases, the technology wave that the patenting wave presages simply never shows up. Sometimes the IP space gets filled with patents for a technology that never takes off. This happens frequently but is rarely reported because it makes for boring news. It is much more fascinating to hear about cases like the smartphone, where the folks that came up with smartphone-related patents many years ago now seem like sages.

Figure 8.8(b) shows an idealized view of how the technology waves and patent waves start to come in sequence once a commercial product hits the market.

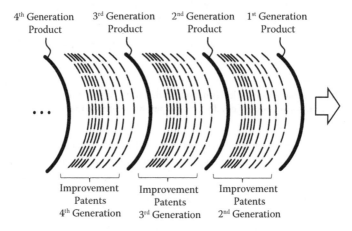

Figure 8.8 (b) Improvement patents precede the corresponding generations' product.

Each generation of the technology and its associated products is anticipated by a wave of improvement patents related to the new commercial implementations. The second-generation smartphone, for example, has a better antenna, a longer lasting battery, and a more durable cover glass than its first-generation counterpart. The third-generation model offers six new buttons for functions most people will never use, better imaging, a faster processor, and more memory. With the fourth-generation model, you get an optical-fiber data port, a brighter screen, an even better antenna for Internet reception calls, a high depth of field, a high-resolution digital camera lens, and so on. At some point, the smartphone will be able to make coffee but not field a phone call.

8.9 THE TRANSITION FROM CORE TO COMMERCIALIZATION PATENTS

Figure 8.9 shows another idealized way of understanding how patents for a given technology evolve from the time the core patents create the IP space to the time a commercial product makes it to market.

Figure 8.9 starts with the creation of an IP space by two core-invention patents (large white circles), which cover most of the space. Next, improvement patents (light-gray circles) on improvement inventions fill in parts of the IP space as the technology matures and a future product starts to take shape.

Notice here that, unlike in our fractal pattern representation of the IP space, patents are now allowed to overlap. This is actually a more accurate picture of what happens, since patents can in fact overlap one another in IP space when viewed along certain dimensions. One can obtain a patent on an invention whose claims "read on" another patent—meaning that practicing the claims of the follow-on patent would infringe the existing patent. This leads us to an important point worth remembering:

> The patentability of an invention is a separate concept from whether practicing the invention would infringe another patent

It can and does happen that an improvement invention that would infringe a core patent if practiced also differs sufficiently from the core patented invention that the follow-on improvement invention is patentable. On the other hand, it can also happen that the core patent renders the improvement invention unpatentable because the two inventions are so close. In the case of the inventions covering CMP polishing pads with windows discussed earlier in this chapter, there are

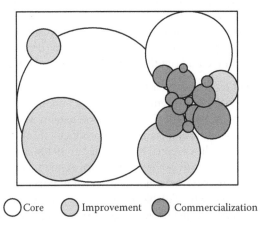

○ Core　● Improvement　● Commercialization

Figure 8.9 Another view of how core, improvement, and commercialization patents fill the IP space.

surely some improvement patents that ***read on*** (i.e., would infringe if the claim were actually practiced) the core patents in this densely packed IP space. This is the case for almost every IP space.

Many patentable improvements are built atop existing patented inventions and, to be implemented, require use of the core invention. Imagine, for example, that someone invents an apparatus for polishing a widget with a laser and that the polishing takes an hour to complete. Now imagine that someone else comes along with an improvement to the apparatus that adds a complex arrangement of critical beam-enhancing components that reduce the polishing time to 1 second.

The patent on the first invention might very well have claims that do not limit the speed at which the polishing occurs. Thus, while the additional components used to form an improvement invention can be sufficiently inventive to yield a patentable invention, the claims of that improvement patent may be subsumed by (i.e., read on) the core patent claim so that the use of that improvement invention can end up infringing the basic apparatus patent.

Referring again to Figure 8.9, at the improvement patent stage, the form the product will ultimately take (if it ever makes it to the market) is still unknown, and the regions where the improvement patents are generated are still based on a good deal of guess work about how the product will ultimately be commercialized.

Later on, when the commercial embodiment of the product is starting to take form, commercialization patents (dark-gray circles) on commercial implementation inventions arrive in select areas of the IP space. These commercial implementation patents make the IP space denser in one or a few areas of intense interest. For example, in the smartphone IP space, the commercialization patents may focus on things like display features and functionality, image-processing capability, longer lasting batteries, more robust cover glass materials, and so on.

Now, the interesting thing about Figure 8.9 is that it can represent the state of an IP space at a time well past the expiration of the core patents and even of some of the major improvement patents. In this scenario, what remains in force are the more recent commercialization patents, with the core patents freely available for everyone to use. This scenario arises when the commercialized product appears long after the core patents were issued. In such cases, the most valuable IP is actually the newly minted and tightly focused commercial implementation IP.

This explains why, in certain cases, a patent portfolio with a couple of core patents and a few first-generation improvement patents isn't enough to cover a commercially viable product. The core and first-generation patents might have expired by the time the market is ready for a commercial product—think 3D televisions and movies, which were actually invented back in the late 1940s and are now suddenly all the rage again. The core patents have long since expired. The patents in the 3D television and movie space now are mainly directed to improvements to the basic technology that rely on newly available technologies, such as fast optical switches, improved polarizers, computing power, image-processing software, and displays.

Someone keen on driving the commercial implementation of a product can, in certain situations, end up capturing the key commercialization patents and creating a roadblock for the industry even though that person didn't invent any of the core patents, or even any of the lower order (i.e., first and second generation) improvement patents. The technology business that owns the amazing set of core patents might be left holding the bag as it watches them expire without ever having followed up on the improvement and commercialization patents needed to stay in the game.

8.10 KNOW THY IP SPACE

The recognized best-practice approach for most serious technology businesses with respect to prior art is to measure and maintain watch on the prior art in the relevant IP spaces to some reasonable degree. It also includes performing a product clearance (also called a freedom to operate analysis) for a new product (at the design stage) to avoid infringement problems. Thus, knowing the IP space in which you live and operate has two main benefits: It increases the probability that your own issued patents will be valid and it reduces the risk of infringing the patents of others.

It is not that difficult to get a first-order measurement of the contours of an IP space. At the very least, a sanity-check search directed to uncovering otherwise conspicuous and extremely relevant references should be performed. It is manifestly embarrassing to make a product and get it to market only to find out there are numerous conspicuous patents in the IP space that present infringement problems.

The various free and searchable databases available on the Internet—including the websites of all the major patent offices in the world—Google Patents, Freepatentsonline, PatentStorm, and a host of others—allow you to perform basic searches for prior art using key words. A number of excellent reference books that teach the basics of how to perform a prior-art search are also available.

8.11 PRIOR ART AND PATENT APPLICATION PREPARATION

It is always much easier to deal with problematic prior art upstream in the patenting process rather than downstream. In nautical terms, it is much easier to navigate around the prior-art boulder that you have spied in the channel up ahead than it is to pull yourself off it once you have run aground. A patent application drafted with the most relevant prior art all charted out will be less subject to a surprise situation requiring last-minute evasive maneuvers, such as trying desperately to narrow the scope of the claims by trying to find useful limitations and embodiments in the specification. Most well-written patent applications include a host of different embodiments and alternative components and configurations that represent fallback positions for the claims. However, sometimes a rejection from the patent office based on a prior art reference you didn't know about can leave you wishing that you had included better distinguishing features in the specification and drawings to amend the claims strategically.

One of the purposes of searching for the most relevant prior art is to craft claims that balance claim scope ("infringeability") with claim validity, as we explored in Chapter 7. To this end, it is helpful to perform your own prior art search and examination of the patent application before it gets submitted to the USPTO and then to dare the examiner to do a better job.

8.12 THE MYTH OF SEEKING THE BROADEST CLAIMS POSSIBLE

A common problem that crops up during the examination of a patent application is that the examiner focuses on prior art that is only marginally relevant, usually because for some reason the most relevant prior art has not been identified. This can happen when the examiner, perhaps not fully understanding the invention, heads off down the wrong path in his or her search and then misapplies the wrong reference or references to the claims. If the patent applicant stands by and, intentionally or innocently, allows this situation to unfold, then the marginally relevant prior art may serve as the witting or unwitting straw man that the applicant can easily shoot down and get around. This gives whatever claims get issued the illusion of credibility.

In some cases, the most relevant references are actually in front of the examiner but are either missed or misunderstood. If the claims have been already tailored to the closest prior art, then a misguided rejection is a minor bump in the road to getting a solid patent issued. If the claims as submitted were overreaching to begin with, then overbroad claims can issue.

Patent owners often do not know much about the legal side of patenting, so they need to rely on a patent attorney to craft and then prosecute (i.e., work with the patent office to obtain) the claims. They usually assume that the attorney will strike the right balance between claim scope, validity, and business value. Yet too many patent attorneys, thinking they are doing their clients a favor, focus exclusively on achieving the broadest possible claims the USPTO will allow, regardless of the state of the prior art and regardless of the business reasons behind the patenting effort.

In some instances, even when solid prior art is submitted with a patent application to the USPTO, attorneys will include claims they know to be overly broad, expecting to narrow the claims only as far as is necessary to overcome whatever rejection the USPTO issues. However, if the USPTO misses the most relevant prior art cited and the examination gets off track, then the overbroad claims will not get properly narrowed.

Now, I hasten to add that, in some cases, an overly broad claim can benefit the patent owner. But because this situation comes with certain risks (which we discuss momentarily), it is up to the **patent owner** and not the patent attorney to determine how much risk to assume. How much, for example, does the patent owner want to risk having his or her broadest claims collapse to an invalid state in the event they are measured? If the client and attorney aren't communicating effectively, then this disconnect can come back to haunt patent owners, who may go away thinking they own more property than they are entitled to.

To understand the problems this situation can create, let's consider a real-estate analogy. Imagine, for example, that you buy a property that includes a house and a field with a nice section of lakefront. You renovate the house at great expense and set up a nice bed-and-breakfast, complete with beach access, a fleet of expensive wooden canoes, and fine rocking chairs that look out over the lake from an expansive porch.

Then one day someone starts building a warehouse in the field between your house and the shoreline. Indignant, you run outside with your deed and point out where your rather faint property line seems to run. Then the warehouse builder pulls out a deed, which has a nice bold survey line and all kinds of official stamps, quite clearly showing that the portion of the property in question is the builder's—not yours.

At this point, you call the lawyer that represented you in the property transaction and ask him what the deal is. The lawyer responds with something like, "Well, you know, we weren't really sure about the actual property boundary with the line being sort of faint and all—but when we registered the deed in the town hall, no one seemed to object to us taking the land all the way down to the lake, and we just assumed you would enjoy having that beautiful waterfront."

Now, wouldn't it have been nice to have known all that before you renovated the house, set up the website, took a bunch of reservations for the summer, and drove your truck all the way to Maine to buy the fancy rocking chairs and expensive canoes?

Getting meaningful and valuable claim protection involves much more than just having the broadest claims possible. When the business angle that the patent owner is supposed to impose on the patenting process is absent, the patenting process tends to revert to an outright IP-space land grab. When, down the road, the client relies on the presumed validity of what are later revealed to be overly broad claims and makes significant investments to sell products and run a business—well, it's not much different from the real-estate analogy.

Some might argue that the narrow dependent claims can be used to save the patent if a broad independent claim is ultimately held invalid. However, this argument presumes that the dependent claims will have the kind of limitations that actually matter. Lots of patents have dependent claims that just add meaningless or trivial limitations to the main independent claims or that go off in the wrong direction. Moreover, in a situation where a patent owner is asserting infringement based on the presumed validity of the broad independent claim and not on the uninfringed narrower dependent claims, the dependent claims aren't going to matter.

Some assertive licensors don't care much about whether their broad claims have a high probability of being found valid if measured because it is their business model to leverage whatever IP they have against as many accused infringers as possible while keeping a straight face. A serious technology business that actually makes and sells products and that does not hope to get involved with patent litigation just for the excitement of it may want to be more thoughtful about matching the scope of its claims to what it has actually invented.

It is worth remembering that the prior art you don't care to look for will be looked for by your competitors, not to mention some very enthusiastic and well-paid opposing counsel and search agencies, should your patent ever become sufficiently valuable and problematic to someone else. A savvy technology business owner will be well informed about the IP space in which his or her business operates and what constitutes a sensible claim scope. Relying on the USPTO or on a patent attorney to make these determinations can lead to inflated claims that aren't matched to the business needs and can lead to serious miscalculations about the worth and breadth of the IP that the patent represents.

8.13 THE FREEDOM TO OPERATE

Here is the question that every technology business that seeks to make and sell a product should ask: Does our product infringe any patents owned by others? That is to say, does the business have the *freedom to operate* to sell the product?

Answering this question properly requires performing the aforementioned product clearance or a freedom-to-operate (FTO) analysis. When a law firm provides that analysis as a legal opinion, is it called an FTO opinion.

Some technology businesses do not bother to perform an FTO analysis. One reason to forgo an FTO analysis is that the results won't affect the business decision to make the product. Not all companies have the luxury of changing course. Many a start-up company, for example, is simply trying to turn a profit while being understaffed and underfunded. If a start-up company sells a total of ten widgets a year for $50 a piece, and its widget infringes five patents of five different companies, it is highly unlikely that any of the patent owners would notice or even care if they did notice. However, if the start-up company sells ten million widgets a year for $1,000 a piece, then you can bet it will get the attention of the previously aloof patent owners.

But by then, the start-up company will be armed with enough money to deal with the infringement issues. Maybe the worst thing that happens is that a licensing agreement is negotiated and disaster is avoided. Or, maybe the start-up company will choose to fight the charge of infringement if it believes the problematic patents can be invalidated. In some cases, it may be better to get the product out the door and into the market to gain traction as fast as possible, regardless of whether your product steps on the IP of others. This happens every day in the technology world.

For more mature technology businesses that would rather not engage in such IP bungee jumping, an FTO analysis provides a way to design the product that minimizes the risk of infringement problems down the road. If necessary, the technology business can negotiate licenses before it commits to manufacturing and selling the product. It is always easier to deal with patent infringement issues before they arise, including by avoiding infringement in the first place. If, on the other hand, the technology business chooses not to perform an FTO, it runs the risk that a patent owner will seek an injunction to prevent the sale of the product altogether. There is no guarantee that the patent owner will simply settle for licensing his or her patent, especially if the product directly competes with the patent owner's product.

The worst reason to skip an FTO analysis is sheer ignorance. The decision not to perform an FTO analysis has to be an *informed* decision. For businesses with products whose underlying technology is part of a densely patented IP space, not taking the time to examine the IP space increases the likelihood that the product will infringe one or more patents, which means an increased likelihood of a patent infringement lawsuit.

NOTES

1. This is not to suggest that people do not try to fill the commercially insignificant regions of IP space. Technology businesses that have high-entropy IP systems do this every day.
2. Round, H. J. 1907. A note on carborundum. *Electrical World* 49:309. See also Shubert, E. Fred. 2006. *Light-emitting diodes,* 2nd ed. New York: Cambridge University Press, chapter 1.
3. See, for example, US Patent Application Publication No. 2006/0267029.
4. Research in Motion (RIM) was able to uncover prior-art documents in Norway published by the Norwegian telecommunications company Telenor that it used in an attempt to invalidate patents being asserted against RIM by NTP.
5. This, of course, is measured in *angstroms.*
6. Optical computing, all optical networking, quadraphonic sound, and bubble memory are a few other high-tech examples that come to mind.

CHAPTER 9

PATENT SYSTEM OPERATIONAL REALITY

9.1 EXISTENTIAL REASONS FOR PATENTING

To understand at the most fundamental level why technology businesses seek patents, we first have to understand why technology businesses exist in the first place. They exist to make a profit for their shareholders. They do not exist so that people can have jobs or have fun at work. Technology businesses are all about making money and, more specifically, are all about making a profit; even more specifically, they are all about maximizing that profit.

To this end, well-run companies implement recognized best practices and procedures in as many aspects of their business as possible. Most companies know how to do this in the well-understood aspects of a business such as marketing, finance, human resources, product development, and manufacturing. The International Organization for Standardization (ISO), for example, sets international standards for manufacturing to motivate the implementation of best practices in manufacturing, and companies that work to meet these standards become "ISO certified."

However, the evolving and increasingly complex role that IP plays in technology businesses and the lack of practical IP knowledge in the business workplace make it challenging for many technology businesses to implement IP best practices effectively. Many technology businesses would be hard-pressed to articulate even the most basic elements of IP best practices. Instead, many technology businesses abdicate the best-practice responsibilities for their IP to law firms and assume that the law firm will take care of everything. While a good law firm will certainly take care of all the legal aspects of the patenting process and be sensitive to business-related issues, it should not be relied on to implement and manage the company's internal business procedures and practices. That is not the law firm's job; it is the company's job.

A technology business that relies on IP to stay competitive and maintain or gain market share has a duty to manage and leverage its IP matters in a manner consistent with the corporate business mandate of maximizing shareholder profit.[1]

9.2 PATENTING OUTSIDE SHANGRI-LA

It is easy to make all kinds of oversimplified and generic statements about what a technology business should do to implement IP best practices for the sake of improving its internal IP system and, in particular, its patenting system. Most books that discuss IP management and IP strategy typically advise technology

Table 9.1 IP Best-Practice Buzz Words and Phrases

• Develop an IP strategy	• Conduct product clearances
• Start mining innovations	• Improve the corporate IP culture
• Adopt an IP vision	• Provide inventor education
• Improve innovation	• Monitor competitor IP
• Perform IP analysis	• Reduce patenting costs
• Generate patent maps	• Improve innovation quality
• Grow the patent portfolio	• Motivate inventors
• Document more innovations	• Generate more innovations

businesses to adopt a long list of new behaviors, implement new procedures and practices, and perform a host of new tasks and activities. A short list of these behaviors, procedures, tasks and activities can be found in Table 9.1.

These are all excellent and useful suggestions that are easy to implement in a technology business that operates in Shangri-La, where time is not an issue, money is not a problem, and people are all well adjusted and live in a state of workplace Nirvana. However, for technology businesses that lie on the outskirts of Shangri-La and beyond, any hopes they have of making changes to an IP system and implementing IP best practices are going to interface kinetically with modern-day operational realities.

So rather than discussing the generalities of IP best practices per se, this chapter focuses on identifying and overcoming specific operational issues that hinder the implementation of some of the more important IP best practices.

9.3 HOW TO KILL INNOVATION

I have lectured on patents and patenting as they relate to technology business at various technology conferences and other venues for over a dozen years. The people who attend my classes and talks range from technicians to CEOs. I usually start with a simple question: "What percentage of patents do you think provides your company with real business value?" Before going on to the next paragraph, if you work for a technology company, I challenge you to answer this question for yourself.

The most frequent response I receive? *Less than 5%.*

Some people say 10%. Others say 1% or 2%. I cannot recall anyone ever saying something as high as 25%. More than a few people have blurted out that most patents are "junk" and "a waste of money."

Whether these sentiments are accurate or the question too vague and unfair to answer is not the point. The point is that many technology workers perceive patents as being mostly worthless. It also shows how technology workers don't see how patents have collective value as patent portfolios. As a result, many technology workers have little or no respect for patents and the patenting process. This is not good news for a technology business that is trying to promote innovation and to motivate its people to participate actively in its patenting system.

But it gets worse. Recall our discussion in Chapter 1 about the adverse impact of überproductivity on innovation. In a typical modern-day technology business, workers are constantly being pressed to do more with less. Technology workers now have smartphones, laptop computers, and a host of high-tech devices that electronically tether them to their jobs. The workday is no longer defined by the amount of time spent at the workplace. People are now accessible 24/7 and, via GPS, are locatable on the planet to within a meter. Many employees are expected to remain accessible after normal working hours. While some technology workers enjoy the flexibility of working from home, this arrangement can end up being a glorified form of house arrest.

I'm sure there are lots of ways to demotivate people and keep them from innovating, but one effective way of achieving this goal is by allowing a "patents are worthless" viewpoint to coexist alongside a managerial philosophy that overworks the people who are supposed to be doing the innovating. In such an atmosphere, it makes no difference whether a company hires a chief IP officer to drive innovation. It does not matter how many IP books executives and managers read about how important IP is to the business. All the happy talk about getting more patents is pointless. Adam Smith was right when he said that IP is created at the grassroots level by common workers. When the employees who actually do the inventing and creative thinking are pushed to work at their theoretical maximum efficiency and don't know or respect what patents can do for the business, there is little hope of effecting meaningful change in a company's patent system.

9.4 STATUS QUO INERTIA

Those technology businesses that skate by with a marginal, mid- to high-entropy IP system eventually are going to have an IP epiphany. This epiphany often comes in the midst of an IP-related crisis or soon thereafter. There are few things that get a company's attention more than a patent infringement lawsuit or watching a competitor patent all the innovations the company could have and should have patented but didn't.

Technology businesses take a huge risk when they manically work only toward short-term goals without regard for longer term IP thinking. It is a bit like not wearing your seat belt when you drive. Everything can be totally fine for a long time. But whatever time was saved and whatever convenience enjoyed by avoiding all that trouble of buckling your seat belt instantly make no sense at all at the moment of impact.

In physics, *inertia* is defined as an object's resistance to being moved. More colloquially, inertia implies a resistance to change. One of the biggest obstacles to overcome in a technology business is *status quo inertia.* It is hard to change things when people think the status quo is just fine. How do you reason with someone who tells you that he or she hasn't been wearing a seat belt for years and is doing just fine? Status quo inertia is a huge problem for technology businesses that don't know they are living in the IP Era of the Information Age. It is hard to convince them that the old rules of IP no longer apply and that the old-school way of handling patents now poses a major risk to their business.

Some technology businesses will go to great lengths to avoid implementing serious IP system changes by instead implementing ridiculous IP system changes. Rather than investing in known best-practice ways to achieve their goals, some technology businesses invent their own fixes for their patenting systems based on their own "how hard can it be?" ideas. In doing so, they ignore tried-and-true IP best practices and end up reinventing the wheel, which usually comes out looking somewhat square.

Here is a classic example of an ad hoc IP practice that is often used by people who don't want to take the time to implement an IP best practice. It begins with someone high up in a small, high-tech company deciding to save the company money by writing up a bunch of provisional patent applications and filing them either himself or herself or through a law firm that has no problem filing the documents without substantial review and analysis. Then, with 1 month left before the 1-year deadline for filing the nonprovisional application, the company asks the same law firm or a different law firm to prepare the regular US application as well as a patent cooperation treaty (PCT) patent application to establish international filing rights.

Of course the do-it-yourself provisional applications do not meet the legal standards for supporting claims that would actually protect the products the company makes or wants to make. This is because the do-it-yourself patent application drafter wrote the applications as white papers that talk about *what* the innovation does and *why* it does it and not as a patent application that actually teaches *how* to make it and use it. All those fancy color plots that show impressive results don't actually explain to anyone how to build a device to obtain those results. Details about how the invention actually is made and used were intentionally left out because they would just "limit the invention."

Meanwhile, the law firm that takes on the nonprovisional and PCT application filings doesn't have the technical expertise to question whether the technical content of the applications is actually enabling. Further, the client puts pressure on the law firm to keep the costs down. Consequently, the law firm ends up preparing and filing dubious nonprovisional patent applications on the cheap based on the poorly prepared provisional patent applications.

The law firm arguably is doing what the client asked it to do. In all likelihood, it has done the best job it could with what was provided and at the negotiated cost. Meanwhile, the company figures the law firm knows what it is doing and assumes that its regular and PCT patent applications are all good. Unfortunately, the result is feeble patent applications that are going to be difficult and expensive to prosecute and may never issue as patents. And if patents issue, sophisticated competitors will quickly see the shortcomings.

This scenario illustrates the classic and all too common ***client/law firm IP disconnect.*** The law firm assumes that the client knows what it is doing and what it wants based on the law firm's own spherical client assumptions and its own lack of technical expertise. Meanwhile, the client throws its information over the wall to the law firm and assumes that the law firm will one day throw patents back over the wall. The "how hard can it be?" mentality that makes everything seem so simple and expeditious at the beginning of the task too often leads to IP disappointment or disaster in the end.

Alas, the IP universe preserves IP karma.

CASE STUDY: The Omnibus Provisional Application

In one real-life example, a small technology business thought it would be a good idea to draft and file an "omnibus" provisional patent application that covered all of the company's IP in one shot. A law firm without the technical expertise in the technology was standing by ready to file the omnibus provisional patent application "as is."

Before the company sent the application to the law firm, however, a nonlawyer consultant with solid technical and IP expertise, as well as business savvy and common sense, reviewed the would-be patent application. The consultant found that the document contained a treasure trove of trade secret information that would have given away key technologies. Moreover, other technologies remained unprotected owing to nonenabling descriptions, which were highly technical and sophisticated with lots of impressive color plots and charts that showed *results* but did not tell anyone how actually to make or use anything.

The consultant convinced the technology company to break the document into separate and cogent enabled innovation disclosures for one set of innovations. Another set of innovations best protected by trade secrets was also documented. The company then followed IP best practices in reviewing and filing appropriate provisional patent applications for the one set of innovations while also properly maintaining the other set of innovations as trade secrets.

9.5 TOP-DOWN AND BOTTOM-UP CHANGE

Many IP experts will tell you that effecting change in a technology business's patenting system can happen only from the top down. Without CEO buy-in, they advise, there can be no reasonable expectation of success. I used to believe this as well—until I started seeing how executives and consultants try to effect those changes under the real-life working conditions of a technology business.

Change from the top down is undoubtedly part of what is required to overcome status quo inertia and effect true IP change. But lasting IP change also has to happen from *the bottom up.* The change from the top has to reflect an understanding of and connection to the folks doing the day-to-day real work. The two ends have to be talking the same language and working toward the same goals.

Bottom-up change is much harder to make than top-down change but is far more effective. Top-down change happens because the boss gives orders or an executive rolls out a program that isn't optional. More often than not, this results in side-long glances from the people in the room that know they've just been handed even more work. Bottom-up change happens only when the technology workers believe something is actually worth doing. As we discuss in greater detail later, the people at the top tend to think in big-picture, low-frequency terms that are short on details. The people at the IP grassroots level tend to think in day-to-day high-frequency terms that, in comparison, are loaded with operational details. And, don't forget that many of them may already think that patents are a waste of time.

The fact is that without relieving technology workers of the insane pressure they are under and without providing them with the means and reasons to work smarter and not just harder on IP, all the top-down pressure in the world won't matter. If the common workers don't come on board of their own volition, no IP anything is going to happen. And without the common workers on board, all the downstream IP best practices are superfluous because no innovation is going to happen.

9.6 BUSINESS–LEGAL–TECHNICAL BALANCE

One IP best practice is to make sure that the patenting system balances the dynamic interplay between its business, legal, and technical components. Almost every patenting system problem can be traced to an imbalance in this key triad. This business–legal–technical (BLT) balance is also essential for dealing with IP problems when they arise.

Many types of physical systems and machines operate over what is called a *dynamic range*, which is the ratio between the smallest to largest possible values of a given operational parameter. For example, a typical LCD television has a range of brightness levels that define its contrast, with a typical dynamic range being 700:1. A typical digital camera has a dynamic range in the neighborhood of 2,048:1, which represents the difference between the smallest and largest brightness values that the camera can resolve. The dynamic range of a stereo describes its ability to reproduce the softest and loudest sounds faithfully.

Implementing and operating a patenting system in a technology business involves executing over a ***large dynamic range of operational details.*** This range spans from the high-frequency, mind-numbing buzz of legal minutia to the low-frequency, ground-shaking rumble of strategic thinking. This concept is illustrated in the graph of the dynamic range of operational details shown in Figure 9.1, which plots the (normalized) attention span of a person versus the operational details in frequency f_D.

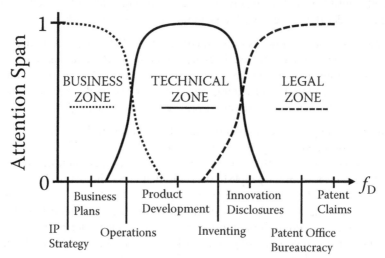

Figure 9.1 A graph of the dynamic range of operational details of a patenting system. It plots the normalized attention span versus the operational details in frequency f_D and shows the business, legal, and technical zones.

In the *business zone,* senior executives such as CEOs, CTOs, and COOs are involved with developing and overseeing business plans, IP strategy, budgets, human resources, operations, and the like. They decide how much time, effort, and resources are to be dedicated to the patenting system; what its ultimate goals are; and what IP space is relevant. These higher-up technology workers tend to think about and operate in big-picture, large-scale, low-frequency terms. Their attention span drops off exponentially as the discussion moves to the higher frequency, day-to-day operational details.

In the *technical zone,* managers, project leaders, inventors, and patent liaisons focus on product manufacturing as well as developing, identifying, extracting, and documenting innovations. Workers in the technical zone work directly with the technology and with practical operational details in mind. These technology workers need to think and operate at midfrequency, practical terms to get the real work done. The attention span of the technical-zone people tends to drop off as the discussion moves either into the big-picture, low-frequency concepts or into the legal, high-frequency intricacies.

In the *legal zone,* patent attorneys and patent agents deal with the minutiae and nuances of the aforementioned patent laws, regulations, procedures, and case law. As I noted earlier, they are charged with serving as native guides to the patent legal system and the USPTO's bureaucracy. Workers in this zone draft and prosecute patent applications as well as scrutinize other companies' patents, in some cases rendering legal opinions about their validity or invalidity and whether a patent is being infringed.

This kind of work inherently involves thinking in and working at microscopic detail scales, with an attention span that stays strong from the mesoscale frequencies all the way up to the highest frequencies. It is not unusual for patent attorneys and patent agents to descend below the atomic scale of detail and into the quantum foam as they spend hours arguing and debating the usage of a single word in a claim or the nuances of a holding in a recently decided patent case. The legal zone is more concerned with arcane details about patent law and the USPTO's byzantine bureaucracy than with big-picture business plans and the company's midlevel operational issues.

This leads us to an important rule:

> **No single zone within the dynamic range of operational details can be allowed to dominate a technology business's patenting system.**

The dominance of any one zone creates an imbalance that inevitably leads to a loss of perspective. As the old saying goes, when you only have a hammer, everything looks like a nail. The people in each of the operational detail zones will by nature want to impose their own viewpoints and prejudices on the operation of the patenting system. If any one zone ends up being dominant, dysfunction and thus patenting system entropy increase.

For example, when lawyers are allowed to dominate a patenting system, the dysfunction can manifest as impressive looking patents that are misaligned with the company's business goals and that lack the highest degree of technical accuracy. The patenting system can become all about getting the patents at the expense of business value and accurate technical information. A key indication of this imbalance is when the attorneys make the process all about "getting claims" without regard to whether the claims will serve a meaningful business purpose. Another indication of an imbalance is when patent applications get filed in culturally interesting but technologically irrelevant countries.

When engineers and managers are allowed to dominate the patenting system, the dysfunction can manifest as a lack of large-scale strategic thinking and small-scale tactical thinking. This can lead to unfocused and marginally relevant if not totally irrelevant patents. This can reach an extreme when inventors drive the process because inventors can tend to think everything they invent is critically important and must be patented.

When high-level executives are allowed to dominate the patenting system, the dysfunction can take the form of the infamous "how hard can it be" mentality, manifesting as commands from on high that are poorly executed because the people doing the detailed work aren't engaged in the process or don't have the necessary training. It can also manifest as patents that, while being in line with the IP strategy, lack the necessary technical information and legal details or that completely ignore the prior art. Such shortcomings can come back to haunt the patent applications during their prosecution or the resulting patents later on in a licensing negotiation or in court.

The dynamic range of operational details must be effectively distributed over the business, technical, and legal zones to ensure the equilibrium and fidelity of the patenting system. This requires teamwork, coordination, and effective communication among the three groups.

9.7 THE CANONICAL PATENTING SYSTEM

Figure 9.2 is a high-level schematic diagram of what can be called a *canonical best-practice patenting process.*

The process is "canonical" because it includes the core elements of all best-practice IP systems. The canonical best-practice patenting process is not new. It is implemented in one form or another in every technology business that seeks patents.

The canonical process includes five key functions: (1) innovation generation, (2) innovation documentation, (3) innovation review, (4) innovation protection, and (5) leveraging. Each function is informed by the *central organizing principles* behind why patents are being pursued, as discussed in Chapter 13. Note that the process is ongoing and cyclical. Invention generation is not a one-time event.

Innovation generation begins with the inventors (i.e., engineers, scientists, and technicians for technical innovations, as well as salespeople, marketing personnel,

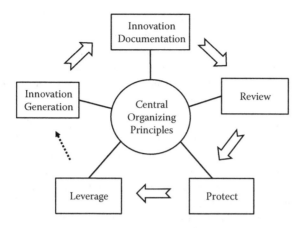

Figure 9.2 The basic functional elements of the canonical patenting process.

and management for business-process innovations), who create innovations in the form of ideas, inventions, and so on in the course of their work. This innovation happens every day in most technology businesses as a natural consequence of the need constantly to improve and refine products and processes.

Innovation documentation is the recording of these innovations in writing. The written documents are often called "invention disclosures" or "records of invention." However, these terms are unfortunate because too many people who come up with innovations consciously or unconsciously construe the term "invention" to mean "patentable invention." Operating on this faulty assumption, they don't bother to document their lowly "innovation." This is a problem because, at this stage of the process, no one should be making decisions about the patentability of an innovation.

Innovation disclosures can be written by the inventor or drafted by an in-house or outsourced IP liaison who interviews the inventor. The innovation review process involves reviewing the innovation disclosures by a "review board" or "review committee" or like group of people. The review process must include input from independent and objective business, legal, and technical people so that decisions are made with the proper BLT balance.

The review committee scrutinizes the innovation from the business, legal, and technical viewpoints and decides on the best disposition of each innovation, such as whether it should be patented, kept as a trade secret, sent back to the inventors for further enablement, put on hold, and so on. Innovation protection is then discussed. Some innovations will be deemed worthy of protecting through patents, while others will be deemed worthy of trade-secret protection. In the latter case, the innovations are logged and the relevant secret information maintained in a manner that affords the information trade-secret status. Other innovations may not be worth protecting as patents or as trade secrets, so they may simply be recorded and the innovation disclosures stored as confidential technical documents.

A common review committee scenario involves an innovation that is certainly patentable and clever but that has no solid connection to the company's business plan. Experienced innovators understand that not all of their innovations will see the light of day as a patent. Rookie innovators can get offended when their innovations are sidelined. In technology businesses that have an award system whereby innovators are given bonuses based on patent application filings, some innovators can get downright testy if their innovation ends up being shelved for good business reasons.

Finally, we arrive at leveraging. Innovations that end up the subject of a patent application can be leveraged in any number of ways, especially if they ultimately issue as patents. For example, some patented inventions may be leveraged defensively to ensure the company has the freedom to practice the invention by preventing others from patenting the same invention. Other patented inventions can be leveraged offensively by either seeking licenses from others or simply preventing others from using the invention—for example, via infringement lawsuits, if necessary. Patents also may be collected as trading chips for cross-licensing agreements to offset the need to take (and thus pay for) straight licenses from others.

Thus, the five key functions in the canonical patenting system (i.e., generate, document, review, protect, and leverage) are guided by the company's IP strategy, which in turn is defined by the central organizing principles (COPs) of the company's business reasons for pursuing patents. The connections between COPs and patent strategy are discussed in Chapter 13.

The last two functions of the canonical patenting process (protecting and leveraging) are actually the easiest to carry out. The protecting part of the process has been the job of patent attorneys for centuries. Once you have an innovation in the form of an invention and that innovation is properly documented, a patent attorney can prepare the patent application, file it, and prosecute it. If the patent issues, then law firms and consulting firms can provide expert assistance with leveraging the patent. If the innovation is better leveraged as a trade secret, this is accomplished in a relatively straightforward manner if well-established best-practice procedures are followed.

That said, patent attorneys are not alchemists. They cannot transform junk innovations into gold. Junk inventions result in junk patents. If a company ends up patenting junk, it has no one to blame but itself. The business-value end of the IP proposition is the ultimate responsibility of the technology business, not the law firm.

9.8 GENERATION AND DOCUMENTATION

The first two functions of the canonical patenting system are innovation generation and innovation documentation. If there is no buy-in from those who consider themselves innovators, then there are not going to be any innovations to document. The downstream review function of the canonical process will devolve into a bunch of people sitting in a room with nothing to do. They will meet every 6 to 9 months instead of monthly. Whatever patenting strategy a company develops is going to be useless if no one has the time or inclination to innovate or, just as bad, document their innovations. Finally, there can be no leverage of IP if there are no innovations to patent or innovations to protect as trade secrets.

As it turns out, the first two functions of the canonical patenting process are by far the hardest to get right. It is not easy to get all the potential innovators in a technology business to start innovating. In the words of Tom Peters, "Creating a corporate capacity for constant innovation is a staggering task."[2] Peters also observes that "constant innovation, from everyone in every function, can only occur if each person is uniquely valued for—and trained to make and paid for—her or his potentially awesome contribution."[3]

Even when a company's technology workers are busy innovating, it can be hard to get them to direct their innovations toward the company's business needs. Likewise, getting people to document their innovations properly so that they can be reviewed and processed can be very difficult. Ultimately, there is little if any difference between the failure of the innovation function and the failure of the documentation function because, at the end of the day, the result is the same: The business has no access to the potential innovation.

Would-be innovators must also have knowledge of which innovations would be considered important to the business. Providing workers with this knowledge is one of the roles of the company's patenting strategy. One way to achieve this consonance is through clearly defined performance goals that effectively communicate the patenting strategy. These goals can alert workers to the kinds of innovations the company needs.

Consequently, for most technology businesses, getting the canonical patenting system to function properly requires: (1) getting the innovators to come up with innovations that actually matter to the business, and (2) getting them (or someone else) to document the innovations. Without these two critical functions, the game is over before it has begun. But when these functions are properly implemented, the rest of the canonical patenting system can be made to run fairly efficiently with reasonable effort.

9.9 NO INNOVATOR LEFT BEHIND

It is not uncommon in a technology business for a small core group of technology workers to end up generating the bulk of the innovations. This phenomenon can create the impression that these few people have some specialized skill that others lack. While it may be true that some of them are more creative than the average person, it is also true that with the proper education and training every technology worker is capable of innovating as part of his or her job.

A technology business that wants to leverage its IP needs to have a "no innovator left behind" policy. This policy ensures that the innovator base is as large as possible. Having a large innovator base essentially kick-starts the front end of the canonical patenting process and gets the innovations flowing downstream. A technology business that has only 10% of its workers innovating and documenting their innovations in all likelihood has a high-entropy patenting system. To keep entropy low, every potential innovator has to be in the game.

On more than one occasion, a technology worker has contacted me (as the outside attorney) to ask where to find the innovation disclosure form for his or her company. It is insane to make inventors traverse an obstacle course just to participate in their own business's patenting system. This leads to a proven IP management theorem:

> **The ability of a technology worker to effectively participate in the patenting system is inversely proportional to the amount of patenting-system inertia he or she must overcome.**

One way to ensure that no technology worker is left behind is to remove all of the obstacles in the way of active participation in the patenting system. For some companies, this starts by not hiding the innovation disclosure forms. More generally, it involves making sure that the patent system of the business is inclusive to all who are available to innovate and not exclusive to a small group that happens to like inventing and knows the ins and outs of the patenting system.

9.10 DOCUMENTING INNOVATIONS IS NOT OPTIONAL

In many technology businesses, innovation happens every day, but the inventors either lack the time to document their innovations properly or have the time but lack the know-how. In some cases, they are too busy writing white papers and technical papers in the belief that these can take the place of a proper innovation disclosure.[4]

Now here are two points to ponder:

> **If a technology business exists to make money and optimize shareholder value, then is it OK for that business to leave large portions of its IP undocumented?**

> **Is it OK to make the documenting of innovations optional?**

The business mandate of optimizing shareholder value means that every employee who comes up with an innovation *must* disclose it in writing to the company. The documenting process, in other words, is *not optional.* The only legitimate business reason for a technology business to leave its innovations undocumented and, ultimately, unleveraged is one that was made consciously at the highest executive level. It cannot be left to chance or accepted as a by-product of a dysfunctional patenting system.

Innovations go undocumented in a technology business usually because of self-inflicted patenting system entropy. This entropy can arise due to the implicit message from management that disclosing innovations to the company in writing is actually an optional exercise. How this implied message is usually conveyed is worthy of its own text box:

> **Management acceptance of allowing innovations to remain undocumented creates the impression that documenting innovations is optional.**

In a technology business that has, say, a hundred people working in jobs that involve innovating, does it make any sense that the same ten people are the only ones innovating? Maybe those ten people enjoy innovating and don't see it as being extra work. But it may also be that the other ninety potential innovators are choosing not to document their innovations. I have been involved in situations at technology businesses where their innovations have generally gone undocumented for *years.* This situation, I have come to learn, is not all that unusual.

Like it or not, innovators do not get to decide unilaterally which innovations matter to the company and which ones don't. As illustrated by the canonical patenting system of Figure 9.2, decisions about the disposition of innovations must be made at the *review* stage of the process so that the proper BLT balance is imposed. This means that *all* innovations must be documented and submitted for review.

This is one of the most important IP best practices. There can be no prefiltering of innovations at the innovator level. With this IP best practice in place, the innovation the innovator had doubts about but submitted anyway can end up being wildly valuable to the business. One of the best problems a technology business can have is finding itself flush with more innovation disclosures than the review board can handle. That is an easy problem to handle.

9.11 DOCUMENT EVERYTHING

Imagine for a moment that you are a manager at a technology business and you are talking to one of your engineers. The engineer has not submitted an innovation disclosure in over a year. You ask that engineer: "Have you not invented anything or made any innovation to any product or process for anything related to your work for over 1 year?" The engineer responds, as many in this situation would, "Well, of course I have, but it's all obvious and trivial stuff."

And there you have it: Adam Smith's common worker filtering innovations at the innovation-generation stage. Unfortunately, decisions about not submitting innovations tend to be made based on infamous IP-related "but I thought that" statements such as "BITT the innovation was unpatentable," or "BITT we don't really use that innovation in our business," or "BITT my improvement wasn't really an innovation."

Once again, employees (and even contractors under certain situations) are not the ones that should be deciding which innovations get submitted to the company. Every innovation must be documented and submitted. This is harder to do than it sounds. For many companies, getting people to create good documentation of even important things like product specifications for products that are already being made is daunting. Heaven help such companies that are trying to get their people to document innovations.

There are two main solutions to the problem of getting all innovations documented in writing so that they can be reviewed and processed:

1. Have a standard innovation disclosure form
2. Provide training on how to use the standard form properly

The form needs to be self-explanatory and self-guided and should not be a burden to fill out. The form also needs to seek very specific information such as the problem being solved, other ways the problem has been solved, the shortcomings of the prior solutions, the gist of the innovation, instructions on how to prepare some simple figures that illustrate the key aspects of the innovation, how the apparatus is configured, possible alternatives and their shortcomings, and so on.

Someone trained in filling out an innovation disclosure form can usually do it in anywhere from 15 minutes to two hours. Anyone who spends a full day writing an innovation disclosure either doesn't know enough about the innovation to be writing it or has a completely mistaken idea of what information is required. The key information that an innovation disclosure must contain is how the innovation differs from what has been done before—that is, how it differs from the known prior art. Details about how to make and use the innovation are obviously important, but at this stage in the game, one can be terse. If the innovation ends up being the subject of a patent application, then the patent attorney who operates in the high-frequency range of operational details will be unleashed and the innovations will be examined and dissected at the molecular and atomic levels of detail.

Requiring workers to use innovation disclosure forms may sound unnecessarily bureaucratic. But there are good reasons for it. The biggest reason is that a good innovation documentation form will elicit precisely the information needed for others to evaluate the innovation from the key business, legal, and technical viewpoints. It is not about the *volume* of information at this point but rather its *quality*. I have read twenty-page-long "innovation disclosures" that could have been better presented as a haiku. One of the hallmarks of a good innovation disclosure is crisp writing that stays on point and answers the questions asked in as few words as possible.

Huge amounts of patenting system entropy get introduced into the process when technology workers decide not to use the standard innovation disclosure form. A classic example is the worker who provides the aforementioned white paper or technical paper in lieu of the innovation disclosure form, appending it to a blank form that reads simply, "see attached." This seemingly expedient approach often leads to hours of wasted time. A white paper that talks at length about *what* an innovation does and *why* it is so great can be utterly useless when it comes to evaluating the innovation from the business, legal, and technical viewpoints that require explanation of *how* to make and use the innovation and *how* it differs from what has been done before. A best-practice patenting system will reject the white paper with a friendly but firm reminder to use proper procedures so that the patenting system can function optimally.

A good innovation disclosure form elicits important business-related information. This includes such things as whether the innovation has been disclosed to anyone outside the company, whether any such disclosure was made under the aegis of a nondisclosure agreement, and whether the innovation has been publicly disclosed, and if so, when.

Technology workers need to understand the importance of sticking with best-practice IP procedures and avoiding cutting corners and submitting spherical approximations to the required innovation disclosure form because it is easier. For the technology worker addicted to submitting white papers and technical papers in lieu of the proper innovation disclosure form, it may be necessary to perform an IP intervention to bring the worker back to IP reality.

9.12 THE IN-SOURCING AND OUTSOURCING OPTIONS

If a company can't implement the IP best practice of having its employees document their innovations in a timely manner, then the documentation process can be in-sourced or outsourced. For example, a patent liaison or a patent engineer, which can be a company employee with a technical background and good listening and writing skills, can be engaged to interview technical workers, collect the necessary information about the particular innovation, and document the innovation in a proper innovation-disclosure format. This in-sourcing approach goes a long way to removing the angst many inventors have about taking the time to write up descriptions of their innovations. Because a patent liaison's evaluation will in large part be based on the number and quality of the innovation documents he or she prepares, the liaison will be highly motivated to document innovations.

Innovation documentation can also be outsourced to an IP consulting firm that specializes in documenting IP, which is one of the relatively new IP-related services available at the IP bazaar. While not inexpensive, for many companies the savings in employee time alone can make it worth the cost. It is not uncommon for a technology business to have a large backlog of undocumented innovations that exist only in the heads of their inventors due to long-term patent system neglect. In this case, outsourcing the innovation documentation process to those skilled in driving and managing this process may be a company's only hope of moving quickly from a mess of undocumented innovations to organized and reviewed innovation disclosures that are ready for patent application preparation or trade-secret disposition.

9.13 THE IP PROJECT MANAGER

Another sticking point that many technology companies face when trying to implement best IP practices is the lack of IP management skills. All too often, someone without any IP experience or interest gets stuck with the job of being the IP manager. The IP responsibilities are usually piled atop all of the person's other job responsibilities. In more than one case I know of, the person assigned to be the "IP manager" thought patents were a waste of time and hated the job.

It can be a much better choice to hire or to outsource this position to a dedicated IP project manager who actually knows what he or she is doing and who actually wants the job. Unfortunately, corporate cheapness too often gets in the way of this sensible option, and the job is dumped on some unwitting and unprepared person who has no idea about how to run a low-entropy patenting system.

9.14 THE NEED FOR PARALLEL PROCESSING

Every technology business has two production lines. The first production line makes the business's tangible products or offers a tangible service, while the second production line makes its intangible IP assets in the form of innovations. In many companies, the second production line is in disrepair and the innovations are inefficiently processed because all the attention is directed to the first production line.

Here is a question that every technology business that relies on patents needs to ask itself:

> **If we manufactured our products like we manufacture our innovations, would we still be in business?**

According to IP best practices, there is no such thing as being too busy to run the second production line. The "we're too busy for IP" attitude arises either because the technology business has been getting along fine without protecting its innovations or because it simply does not know how to run both production lines in parallel. The excuse that "we would do both, but we are too busy and must prioritize" is not acceptable in the IP Era of the Information Age.

Any business that runs its tangible production line at the expense of its intangible production line risks its tangible production line getting shut down by a competitor's intangible asset created and processed on the competitor's IP best-practice intangible production line. Ultimately, the only acceptable answer for why a second production line is not operating is "because we don't want it to." As long as the company owners and/or shareholders are OK with this strategy, there is no problem.

For a technology business that wants to run its second production line efficiently, IP best practices are ready and waiting. This book and others that are more focused on the specifics of IP management can certainly help at the low-frequency, big-picture level. But they can get a technology business only so far down the road. Every business is different. Those who need serious help getting their intangible production line up and running are going to need to address the high-frequency operational details and their own operational realities. Fortunately, there are many good IP and business consulting companies over at the IP bazaar that can get into the details and facts of the particular situation. There are also a lot of bad and overpriced ones, so be careful.

9.15 A DEEPER LOOK INTO THE INNOVATION REVIEW PROCESS

Another IP best practice is to have a thorough and efficient innovation review process, so let's take a moment to delve into this aspect of the canonical patenting system. As discussed before, the review process involves scrutinizing a documented innovation from the business, legal, and technical viewpoints. As we discussed earlier, people who work in the business, technical, and legal zones operate in different dynamic ranges of operational details. A technology company's patenting system must resist being dominated by any one zone or risk an imbalance that introduces dysfunction into the system. Dysfunction leads to bad decision making, which is often followed by a cascade of unfortunate consequences.

Technical workers who submit innovations to a technology business are by definition qualified to speak about the technical aspects of their innovations. However, they can lack the objectivity necessary to make a balanced assessment of the business contribution of their innovations. Some believe that if they invented it, then it must be patented immediately.

Patent attorneys are well qualified to speak to the various legal requirements for the patentability of an innovation. However, they are not the best qualified parties to address the innovation's business value. Also, while an attorney may be technically competent, the attorney is not usually as technically competent as the innovators. The businesspeople, on the other hand, are well attuned to what innovations are needed for the business, but they may not understand the technical nature of the innovation. They also typically do not know enough about patent law to begin to address the legal issues that may be involved.

The BLT discussions during the review process are often fascinating because of the dynamic interplay between the BLT forces. In one meeting in which I participated as a noninventor technical participant, I watched in disbelief as the attorney called the innovation "obvious" and killed it despite pushback from the business and technical people. I have seen businesspeople declare that an innovation is great, only to have the noninventor technical participants send the inventors back to work because there was a suspicious lack of enablement. When I worked at IBM as an optical engineer and inventor, the review committee sent me back to the drawing board for more work on enabling my innovation, a decision that was completely justified and made for a better patent.[5] I have also seen the kibosh put on great innovations that would have made valuable patents because the patent attorney was able to elicit information from the innovators and others that indicated a public disclosure of the invention by the inventors exceeded the 1-year time limit and rendered the invention unpatentable.

At the end of the day, the question we must ask of an innovation disclosure is this: "What value will a patent on the disclosed innovation bring to the business?" If the potential business value of a patent on the innovation is low or nonexistent, then the inquiry is over. If the potential business value is high, then the innovation is scrutinized from a technical viewpoint to make sure there is sufficient enablement to support both suitably broad claims and how the innovation might be expanded and generalized.

The innovation is also scrutinized from the legal viewpoint to see, for example, whether there have been any on-sale activities, public disclosures (e.g., at a trade show), upcoming disclosures under nondisclosure agreements, issues relating to prior art, and so on. Budgetary considerations are also weighed. Even innovations that could prove useful to a business might need to be passed over if the IP budget is limited and there are other, more important innovations in the pipeline that need protecting.

It is important to remember that the review process can also affect the form an innovation ultimately takes. Some of the most important discussions about innovations that I have as an attorney take place as conference calls with the innovators, technical participants, and businesspeople all on the line. We share how we each see the innovation and its potential problems based on our respective expertise. It is amazing how the innovation can evolve as each viewpoint is pressed over the course of the discussion. Often an innovation will appear completely different at the end of the phone call than it did at the beginning. Just as often, what starts out as an "obvious" innovation becomes increasingly less obvious as the issues are more deeply explored and the justification for the innovation is clarified.

On more than one occasion, I have heard an innovator say, after reading a draft of the patent application, "I almost didn't submit this innovation because I thought it was not important, but when I read the patent application, I better appreciated its value and saw that it was not as obvious as I had initially thought." This is often a direct result of the BLT conversations that take place during the review process.

9.16 THE PATENT APPLICATION REVIEW PROCESS

Yet another IP best practice is training inventors on how to review draft patent applications. Inventors need to appreciate that their work does not end with the act of innovation. The inventor needs to be available to perform this key task as well as to sign off on the various patent-related documents that need to be filed with the patent application. While inventors are somewhat justified in viewing the act of reviewing a patent application as drudgery, it is undoubtedly a critical part of the "protect" function of the canonical patenting process. Fortunately, a few things that inventors can keep in mind when reviewing patent applications can make the job easier and more endurable.

The first is that when there are multiple inventors, the review process needs to be coordinated in series among the different inventors. This sounds obvious, but I cannot tell you how many times a patent application was sent out to multiple inventors for review and each inventor reviewed and edited the application without regard to the other inventors. This uncoordinated parallel review process creates multiple drafts of patent applications that are guaranteed to have conflicting information. The patent review process has to be carried out in series on a single "active" document and not in parallel on multiple documents. This usually requires one inventor to take the lead and coordinate the review process with his or her co-inventors.

Second, inventors must keep in mind that the "detailed description" section of a patent application is exactly that—it is supposed to disclose details about how to make and use the innovation. This means not only describing the innovation in general terms but also providing example embodiments, alternative components, different functionalities, and so on. I often hear inventors say, "BITT if we disclose that, we will be limited to that type of device."

This is one of the biggest misconceptions inventors have about patent applications and patent documents. For an innovation to be enabled, the specifics must be disclosed. Furthermore, the patent attorney may need to rely on such specifics if and when the claims need amending. The claims section, on the other hand, can be highly generalized when compared to the detailed description section. You can think of the detailed description section and the claims section as representing the yin and yang of the patent or patent application (Figure 9.3).

Here is why it is OK to put details into the detailed description section of a patent application:

> **The claims section—not the detailed description section—defines the scope of patent coverage.**

Figure 9.3 The yin and yang roles of the detailed description and claims sections of a patent.

9.17 TYPES OF PATENT CLAIMS

Those who review patent applications need to understand something about patent claims. Many inventors who are asked to review the application for their patent end up either ignoring the claims or trying to edit them without any idea about claims-drafting rules or why certain types of claiming language and structure are being employed.

There are two main types of claims: independent and dependent. The independent claims are ones that stand alone and are the broadest, while the dependent claims refer back to an independent claim and add limitations to it. Thus, dependent claims are by definition narrower than the independent claims to which they refer.

To see how these two kinds of claims are used in practice and the roles they play, let's consider an independent claim 1 and four dependent claims (2 through 5) for a hypothetical laser-based mousetrap (the boldface in the dependent claims is added for emphasis and is explained later):

- Independent claim:
  1. A laser mouse trap apparatus for trapping a mouse, comprising:
     a. a container having an open end and an interior sized to contain the mouse;
     b. a laser arranged to direct a laser beam into the container interior, the laser beam having at least enough energy to stun the mouse; and
     c. a sensor adapted to detect the mouse within the interior and activate the laser when the mouse is sensed.

- Dependent claims (because they depend from claim 1):
  2. The mouse trap apparatus of claim 1, **wherein** the laser is a CO_2 laser.

  3. The mouse trap apparatus of claim 1, **further comprising** a door that closes the open end when the laser beam is emitted by the laser.

  4. The mouse trap apparatus of claim 1, **wherein** the laser beam has sufficient energy to vaporize the mouse.

  5. The mouse trap apparatus of claim 1, **further comprising** a source of sleeping gas that is triggered by the sensor to release the sleeping gas to put the mouse to sleep.

The independent claim 1 includes the word "laser," while dependent claim 2 limits the laser to be a CO_2 laser. Including the specific type of laser in the dependent claim serves to inoculate the independent claim from a narrower interpretation than the word "laser" really denotes—for example, "CO_2 laser." This is because the courts are reluctant to read limitations into a claim, especially if the language at issue is included later on in a dependent claim.

This is such an important concept for inventors to grasp that it deserves its own text box:

> **Dependent claims can be used to define a component more narrowly in the independent claim in order to inoculate the independent claim against a narrow interpretation of that component.**

This type of dependent claiming often uses the word "wherein" (as in claims 2 and 4 in the preceding list) and requires that the detailed description include specific examples that support the added claim limitations. As a general rule, inventors should err on the side of providing too many details about the innovation and let the patent attorney decide when enough information is enough.

Dependent claims also serve another important purpose—namely, adding additional components, as opposed to simply narrowing an existing component. This kind of increasingly limited claim scope (which typically includes the phrase "further comprising" or "further including," as in claims 3 and 5 in the preceding list) can raise a claim's chances of collapsing to a valid state relative to the prior art if it is ever measured.

Of course, as we now know that the more limited in scope the claim is, the lower its probability of catching potential infringers becomes (i.e., its "infringeability" decreases). Dependent claims can thus be used to enhance claim validity, albeit at the expense of infringeability.

In many patents, the independent claim is intentionally drafted as a no-shame claim and the dependent claims are drafted to dial back the claim scope increasingly in the hope that at some point the claims will switch from being invalid to valid. As is often the case, however, the wildly broad and invalid claims capture an infringer while the more limited and thus more likely valid dependent claims do not.

9.18 CLAIM CONSTRUCTION

Rest assured that a company accused of patent infringement is going to try to get the problematic claim construed in a manner that is most favorable to it. It will do its level best to have select terms in the claim interpreted in a way that helps avoid infringement and will try to sneak in or read in limitations that aren't really there. Now, before you start thinking of an infringer as the bad guy, remember that it could be your company.

To the extent that the description of the invention in the patent application is vague (i.e., lacks the appropriate details, including the definitions of terms, etc.), the terms and phrases in a claim will be subject to debate and interpretation. Contrary to what many people think, the details included in the description of the invention actually help to eliminate ambiguity in claim terms and make the claims less susceptible to efforts to gerrymander their boundaries.

Here is an all-too-common scenario worth avoiding: A group of inventors is wary about providing details about their innovation because they think the details will limit the patent coverage. Their view holds sway, and the application is filed with a highly generalized and vague detailed description. Then the patent office rejects the claims based on prior art. Now the claims need to be narrowed. The inventors then suggest limitations to the claims based on information available before the application was filed but that was purposely left out of the patent application due to inventor angst about having a "narrow" patent.[6] Unfortunately, at this point in the game, there is no support in the application for the needed limitations, which cannot be added after the fact. So now the claims cannot be narrowed through amendment in the face of the prior art.

Once inventors appreciate the different roles of the detailed description section and the claims section, they should no longer be reluctant to provide detailed information about different alternatives to system components, ranges of operation of select parameters, performance tolerances, example designs, and like details.[7]

9.19 BUSINESS REVIEW OF PATENT APPLICATIONS

In many instances, once the decision to prepare and file a patent application has been made, the focus turns to the technical and legal aspects of the innovation, with the inventor or inventors representing the technical side and the patent attorney representing the legal side. The inventors and the patent attorney work together to prepare the patent application. Usually, if the inventors are happy with the application, it is considered ready to be filed.

An IP best practice is to scrutinize the patent application in terms of its business value, either as it is being drafted or after a draft application is close to being finalized. As we now better appreciate, the patent attorney thinks in the highest frequencies of detail, followed closely by the inventors. This combination of relatively high-frequency viewpoints might skew the process away from business common sense and yield claims with less than optimal business value. Another way of saying this is that it is easy for the inventors and the attorney to lose their collective minds from working too hard at the high-frequency details.

Conducting a business review of a patent application before it is filed can restore some BLT balance (if not sanity) to the process in case it has shifted too far in the legal and technical directions. Put differently, it is unwise to assume that the BLT balance achieved when the review committee decided to pursue a patent application will be maintained throughout the patent application process without any business input along the way. It can't hurt to have a business review of a patent application before it is submitted to the USPTO just to make sure that the claims are properly focused on meeting the needs of the business.

NOTES

1. Poltorak, A. I., and P. J. Lerner. 2000. Corporate officers and directors can be liable for mismanaging intellectual property. *Patent Strategy and Management* 1 (1 and 2).
2. Peters, T. *Thriving on chaos: Handbook for a management revolution,* 337. New York: Harper and Row.
3. Ibid.
4. "BITT I could just submit my technical paper instead of the innovation disclosure form even though the latter asks for lots of specific information that is not in my technical paper."
5. See US Patent No. 5,680,588.
6. "BITT limiting information can be added after the application is filed since we are narrowing the innovation, not expanding it."
7. By way of example, take a look at how much detail is included in US Patent No. 5,680,688 and then see how broadly the first claim is drafted. This is typical of the contrast between the specificity of the detailed description section and the generality of the broadest claims in the claims section.

CHAPTER 10

THAT'S OBVIOUSNESS!

10.1 THE REQUIREMENT FILTERS REVISITED

In Chapter 2 we introduced the main requirements that a US patent application covering an invention needs to satisfy to issue as a valid US patent. We observed that the requirements operate as a series of filters through which a patent application must pass if the USPTO is to issue the application as a patent. It should come as no surprise that some of the filters are easier to apply than others.

Of all the requirement filters, the nonobviousness filter is arguably the hardest filter to apply because it involves combining information from different sources to assess whether a person of ordinary skill in the art of the invention could have or would have readily come up with the same invention under the same circumstances given the *entire body of available prior art*. That is to say, the person of ordinary skill in the art is assumed to have *extraordinary* understanding of the prior art—and, in fact, is assumed to be omniscient in this regard. The analysis is far more expansive than the analysis required to establish a lack of novelty, which needs to be based on a single prior-art reference or source (e.g., an existing product). The obviousness analysis is also much more subjective than the analysis for the other requirements.

Because rejections based on the nonobviousness requirement are a common form of claim rejection and often most difficult to deal with, it is worth drilling down into the details to better understand this particular patentability requirement.

10.2 THE HISTORY OF OBVIOUSNESS

The nation's first patent act was passed by Congress in 1790 and it required that a patentable invention be something "not known before or used"—that is, novel. It did not include a requirement that an invention be "nonobvious." However, it soon become apparent that the novelty requirement alone was insufficient to ensure that patents would be granted only for inventions that offered something more than either trivial improvements to existing inventions or straightforward combinations of known elements, parts, and machinery that functioned in expected ways.

The situation reached a head in 1851 when the US Supreme Court, in *Hotchkiss v. Greenwood*, 11 How. 248 (1851), weighed in on the nonobviousness problem for a patent on a method of making knobs (e.g., doorknobs). The method utilized clay or porcelain but otherwise employed the same methods used to make wood and metal knobs.[1] The Supreme Court ended up invalidating the patent for being "destitute of ingenuity or invention," calling it "the work of a skillful mechanic, not that of the inventor."

Subsequent court cases affirmed a fairly straightforward nonobviousness standard, with the Supreme Court stating in a case in 1950 that a "patent for a combination that only unites old elements with no change in their respective functions...obviously withdraws what is already known into the field of its monopoly and diminishes the resources available to skillful men."[2] This comment elaborates on the point we made back in Chapter 8 about patents that occupy forbidden regions of IP space where the claimed invention should be considered obvious. Such patents unfairly limit the ability of others to use technologies that should be freely available.

In 1952, the nonobviousness requirement was incorporated into the statutory patent law in 35 USC § 103, which reads in part as follows:

> *A patent may not be obtained...if the differences between the subject matter sought to be patented and the prior art are such that the subject matter as a whole would have been obvious at the time the invention was made to a person having ordinary skill in the art to which said subject matter pertains.*

Not long after, in 1966, the Supreme Court heard three patent cases that hinged on the nonobviousness requirement and articulated a nonobviousness analysis in *Graham v. John Deere* 383 US 1, 17 (1966). In the *Graham* case, the Court looked to its analysis in the *Hotchkiss* doorknob case and set forth three main factors (known as the "*Graham* factors") that must be considered when determining obviousness:

1. The scope and content of the prior art
2. The differences between the prior art and the claimed invention
3. The level of ordinary skill in the pertinent art

Under the *Graham* factors, an invention comprising a combination of existing elements (which virtually all inventions do) must show:

- Some change in [the elements'] respective function; or
- A new and different function; or
- An effect greater than the sum of the several effects taken separately.[3]

A nonobviousness analysis based on these factors was and is intended to enable a "broad inquiry" for assessing patentability and not to limit the analysis to a particular line of reasoning.

Further to this point, the Supreme Court in *Graham* also stated that the nonobviousness inquiry could include, where relevant, secondary considerations such as commercial success, the meeting of long-felt but unsolved needs, failure of others to solve the problem the invention solves, and so on— to shed light on the question of nonobviousness. The Court realized that even with the *Graham* factors, questions of nonobviousness would be difficult to address uniformly in all situations and that the patent laws would need to develop over time as more and more cases were decided based on the *Graham* factors.

10.3 TWO SUPREME COURT OBVIOUSNESS DECISIONS

Two Supreme Court cases provide good examples of how the Court measured two different patents, with one collapsing to an invalid state and the other to a valid state. The first case is *Anderson's-Black Rock, Inc. v. Pavement Salvage Co.* 396 US 57 (1969). The patented invention before the Court was a device that combined two existing elements: a radiant-heat burner and a paving machine. The machine was used to heat asphalt as it was being spread to eliminate the known problem of "cold joints" forming between adjacent asphalt layers. The Court concluded that the device as claimed did not create a new synergy between these existing elements because the radiant-heat burner functioned just as a burner was expected to, as did the paving machine. In other words, in combination, the two claimed components did no more than they would in separate, sequential operation.

The Court observed that "while the combination of old elements performed a useful function, it added nothing to the nature and quality of the radiant-heat burner already patented." As an attorney working on another later case (i.e., the *KSR* case, discussed later) observed of this case, all the inventor had done was to take a burner that functioned as a burner and a spreader that functioned as a spreader and put them together without any unexpected result or function.

In *United States v. Adams,* 383 US 39, 40 (1966), a companion case to *Graham,* the Court considered the obviousness of a patented invention for a wet battery. The battery differed from prior-art batteries in two main ways: It contained water rather than the acids conventionally employed in storage batteries, and its electrodes were magnesium and cuprous chloride rather than zinc and silver chloride. One could say that the claimed battery invention was similar to the prior art and simply substituted different known elements for the existing elements. However, the Court noted that, in such a case, the question must be whether the combination does more than yield a predictable result.

As it turned out, the prior art warned of risks involved in using the types of electrodes Adams employed in the design of his battery. The Supreme Court therefore ruled that the prior art "taught away" from the combination of elements that Adams employed in his battery. The fact that these elements worked together in an unexpected and fruitful manner resulted in the Court's finding that the claimed battery was not obvious. Thus, the nonobviousness inquiry includes the principle that when the prior art teaches away from the combination of known elements under review, the discovery of a successful means of combining them is more likely to be nonobvious.

10.4 THE *KSR* CASE

In 2007, the US Supreme Court once again took up the question of nonobviousness in the case *KSR International Co. v. Teleflex Inc., et al.* 550 US 398 (2007). Teleflex, Incorporated ("Teleflex"), owned US Patent No. 6,237,656, invented by

Steven J. Engelgau, on an invention directed to connecting an adjustable vehicle-control pedal to an electronic throttle control.

10.4.1 The patented invention

Figure 10.1 shows Figure 1 and Figure 3 of the Engelgau patent. Referring to these patent figures (and paraphrasing the language set forth in the Engelgau patent), a vehicle, 10, has a control pedal apparatus, 12, that includes a pedal arm, 14, that can be adjusted in fore and aft directions with respect to the vehicle by a driver, 16. This adjustment capability allows the pedal arm, 14, to be positioned to accommodate drivers, 16, of various heights. A pivot, 24, pivotally supports an adjustable pedal assembly, 22, with respect to a vehicle structure, 20, and defines a pivot axis. An electronic throttle-control mechanism, 28, is attached to the vehicle structure, 20, for controlling an engine throttle, 30.

The electronic throttle-control mechanism, 28, is responsive to the pivot, 24, and provides a signal, 32, that corresponds to a pedal-arm position as the pedal arm, 14,

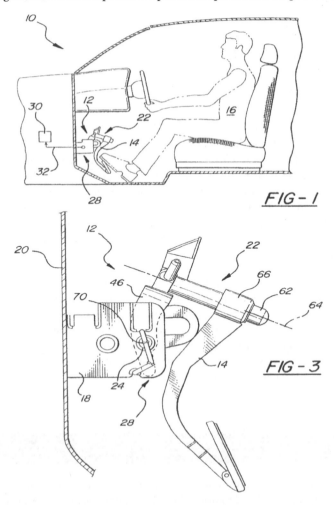

Figure 10.1 Figures 1 and 3 from the Engelgau patent.

pivots about the pivot axis between rest and applied positions. Thus, the signal 32 will vary as the pedal arm 14 moves from the rest position to the applied position. The electronic throttle-control mechanism 28 can be any of various electronic throttle-control mechanisms known in the art, such as the one described in US Patent No. 5,819,593.

Claim 4 of the Engelgau patent, which we saw back in Chapter 7, is reproduced here again:

> **4.** A vehicle control pedal apparatus (**12**) comprising:
>
> a support (**18**) adapted to be mounted to a vehicle structure (**20**);
>
> an adjustable pedal assembly (**22**) having a pedal arm (**14**) moveable in force and aft directions with respect to said support (**18**);
>
> a pivot (**24**) for pivotally supporting said adjustable pedal assembly (**22**) with respect to said support (**18**) and defining a pivot axis (**26**); and
>
> an electronic control (**28**) attached to said support (**18**) for controlling a vehicle system;
>
> said apparatus (**12**) characterized by said electronic control (**28**) being responsive to said pivot (**24**) for providing a signal (**32**) that corresponds to pedal arm position as said pedal arm (**14**) pivots about said pivot axis (**26**) between rest and applied positions wherein the position of said pivot (**24**) remains constant while said pedal arm (**14**) moves in fore and aft directions with respect to said pivot (**24**).

Both the district court and the Supreme Court boiled down the "patentese" of claim 4 to the following clear and understandable description of the claimed invention:

> *A position-adjustable pedal assembly with an electronic pedal position sensor attached to the support member of the pedal assembly. Attaching the sensor to the support member allows the sensor to remain in a fixed position while the driver adjusts the pedal.*

KSR International (hereinafter "KSR") is a Canadian company that manufactures and supplies auto parts, including pedal systems. Ford Motor Company hired KSR in 1998 to supply an adjustable pedal system for various lines of automobiles with cable-actuated throttle controls. KSR developed an adjustable mechanical pedal for Ford and obtained US Patent No. 6,151,976 for the design. In 2000, General Motors Corporation chose KSR to supply adjustable pedal systems for its Chevrolet and GMC light trucks that used computer-controlled engines. KSR merely took its patented design and added a modular sensor.

Teleflex found out about the KSR product and sued KSR, alleging that the KSR product with the modular sensor infringed claim 4 of the Engelgau patent. KSR countered by alleging that claim 4 of the Engelgau patent was invalid for obviousness.

10.4.2 The IP space context

As we discussed in Chapter 8, in cases where the given technology is already embodied in a commercial product, the IP space is almost always going to include a substantial amount of close prior art. In the *KSR* case, the Court looked at the IP space at the time when the Engelgau patent was issued and found that, since the 1970s, plenty of prior art had been directed to pedals designed to have an adjustable location in the footwell of a vehicle for the purpose of accommodating drivers of different sizes.

Two patents relating to adjustable pedals that existed in the IP space at the time at which the Engelgau patent issued were US Patent Nos. 5,010,782 (inventor: Yasushi Asano) and 5,460,061 (inventor: Harry L. Redding). The Asano patent discloses a support structure for housing a pedal so that even when the pedal location is adjusted relative to the driver's size, one of the pedal's pivot points stays fixed. In the Asano pedal, the force necessary to push the pedal down stays the same regardless of adjustments to the pedal's location. Figure 10.2 is Figure 1 of the Asano patent.

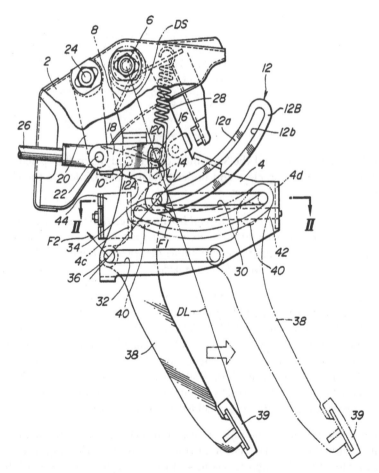

Figure 10.2 Figure 1 from the Asano patent.

Interestingly, the Asano patent was not considered in the examination of the Engelgau patent application even though its presence in the IP space was not exactly inconspicuous. One might be tempted to conclude that Teleflex purposely hid its knowledge of the Asano patent. However, given that this point did not appear to be alleged at the trial, it is fair to assume that Teleflex did not know about the Asano patent. This circumstance, by the way, serves as a good reminder that sometimes the most relevant prior art remains out of sight for mysterious reasons.

The Redding patent also discloses an adjustable pedal, but in this case the pedal has a different type of sliding mechanism and both the pedal and the pivot point are adjustable. The related IP subspace of electronic sensors and computer-controlled throttles also has its own prior art. For example, US Patent No. 5,241,936 (inventor: Jay D. Byler) teaches that it is preferable to detect the pedal's position in the pedal assembly and not in the engine. The Byler patent further discloses a pedal with an electronic sensor on a pivot point in the pedal assembly. Also, US Patent No. 5,063,811 (inventor: Ronald A. Smith) teaches that to prevent the wires connecting the sensor to the computer from chafing and wearing out and to avoid grime and damage from the driver's foot, an electronic sensor should be put on a fixed part of the pedal assembly rather than in or on the pedal's footpad.

Further, US Patent No. 5,385,068 (inventor: James E. White) discloses a type of self-contained modular sensor. A modular sensor is designed independently of a given pedal so that it can be taken off the shelf and attached to mechanical pedals of various sorts, enabling the pedals to be used in automobiles with computer-controlled throttles. In 1994, Chevrolet manufactured a line of trucks using modular sensors that were attached to the pedal support bracket, adjacent to the pedal, and engaged with the pivot shaft about which the pedal rotates in operation.

The IP space at the time the Engelgau patent issued also included prior-art patents directed to the placement of electronic sensors on adjustable pedals. For example, US Patent No. 5,819,593 (inventor: Christopher J. Rixon) discloses an adjustable pedal assembly with an electronic sensor for detecting the pedal's position. In the Rixon patent, the sensor is located in the pedal footpad.

It needs to be emphasized here that the obviousness analysis assumes that one skilled in the art is omniscient when it comes to the prior art. That is to say, the obviousness analysis looks at all the art that existed prior to the invention, not just what was known to the inventor.

10.4.3 The path to the Supreme Court

The *KSR* case was first heard at the district-court level, and the district court found claim 4 to be obvious and thus invalid. Teleflex appealed this decision to the Court of Appeals for the Federal Circuit (CAFC), which reversed and found claim 4 to be nonobvious and valid. KSR then appealed the CAFC decision to the Supreme Court.

Now, the Supreme Court can handle only so many cases in each of its sessions and thus agrees to hear a case only when it thinks that the lower court's ruling is worthy of review, usually with a larger picture in mind regarding the state of the law and not just to decide a specific case based on specific facts. Clearly, the Supreme Court thought something was not quite right with the *KSR* decision at the CAFC and decided to hear the appeal. The Supreme Court's decision to take the case had an interesting effect on the CAFC, which almost immediately figured out that it must not be doing things exactly right and revised the way it analyzed questions of obviousness. So, in this case, the very act of the Supreme Court taking the case caused a change in behavior at the CAFC level.

The CAFC had reversed the district court's ruling in *KSR* by applying a fairly narrow test for nonobviousness known as the "teaching suggestion motivation," or TSM, test. The TSM test was developed over the years, gaining traction in the 1960s, and used with increasing regularity since then until it became essentially the main test for assessing nonobviousness at the CAFC; the USPTO also uses the TSM test in examining patents. The TSM test requires, for a claimed invention to be found obvious, that there be some teaching, suggestion, or motivation in the prior art to combine the elements the invention combined.

The philosophy underlying the TSM test is that essentially all inventions consist of previously known elements and that all inventions can be made to look obvious in hindsight. The TSM test seeks to prevent an examiner or a court from using the claim as a template to go looking for the different claim elements in different prior-art references and piecing them together to arrive at the claimed invention. The *KSR* case, however, challenged the narrowing effect that the TSM test had on the scope of the nonobviousness analysis. In particular, KSR argued that the test ignored previous Supreme Court precedent cases that called for a broad and expansive analysis based on the *Graham* factors as discussed earlier.

10.4.4 Claim 4 of the Engelgau patent

In the oral argument before the Court, KSR's attorney was asked why KSR was seeking to invalidate only independent claim 4 of the Engelgau patent. The attorney's response was as follows:

> *[Teleflex has] not asserted claims 1, 2 and 3 in this case because those claims don't describe anything remotely like [KSR's] pedals. They limited their claim to claim 4 because only by claiming this enormous verbal abstraction that is claim 4 can they make a colorable claim of patent infringement against [KSR] in this case.*

So, in other words, claims 1, 2, and 3 included a fair number of limitations that prevented these claims from encompassing the KSR product. These claims belonged to zone 3, where claims are drafted to have a better chance of collapsing to a valid state if measured but at the expense of capturing infringers, such as KSR. Because no one was being accused of infringing claims 1, 2, or 3, there was no reason to measure them.

Independent claim 4, on the other hand, was a zone 2 offensive claim. Despite the patentese, claim 4 was drafted so broadly that it had a high probability of ensnaring an infringer (in this case, KSR) as well as a high probability of collapsing to an invalid state if measured.

KSR dragged claim 4 all the way to the Supreme Court to get the ultimate legal measurement because it believed, despite the CAFC measurement, that the claim

4 wavefunction would collapse to an invalid state when it was properly measured. The attorney for KSR put it more elegantly when he stated that claim 4 "sweeps so broadly, so much broader than what the applicant in fact invented, that it sweeps in obvious manifestations."

Naturally, Teleflex argued that its claim 4 was valid. At one point in the oral argument before the Supreme Court, Teleflex's attorney cited an expert opinion that described the complexity of the claimed invention in an effort to show that the claim was nonobvious. This prompted the following exchange between Chief Justice Roberts and Teleflex's attorney:

> *Chief Justice Roberts: Who do you get to be an expert to tell you something's not obvious?*
> *Mr. Goldstein: You get...*
> *Chief Justice Roberts: I mean, the least insightful person you can find? (Laughter.)*
> *Mr. Goldstein: Mr. Chief Justice, we got a PhD and somebody who had worked in pedal design for 25 years.*
> *Chief Justice Roberts: Exactly.*

This exchange raises a good point, which is that both sides in a dispute over obviousness are going to find an expert in the field that will give an opinion about the invention that supports the view their side is espousing. While these opinions are often helpful, it is equally true that they can obscure the reality of the invention. While a court will give credence to such opinions where credence is due, at times, the opinions defy logic and common sense and are dismissed as self-serving nonsense. Chief Justice Roberts had little tolerance for a fancy opinion that flew in the face of what he saw as a plainly obvious invention.

Justice Breyer's view of the invention during the oral argument was also quite telling and lends credence to the characterization of the Court as fairly commonsensical in its evaluation of obviousness.

Here is Justice Breyer's impression of the Engelgau invention:

> *You look at that thing, you think what this genius did, and I don't doubt that he's a genius, is there are wheels that turn around. And the wheels turn around to a fixed proportion to when you make the accelerator go up and down. Now I think since high school a person has known that if you have three parts in a machine and they each move in a fixed ratio one to the other, you can measure the speed of any part by attaching a device to any other as long as you know these elementary mathematics. I suppose it wasn't Mr. Engelgau, it was probably Archimedes that figured that one out. So he simply looks to something that moves, and he sticks a sticker on it. Now to me, I grant you I'm not an expert, but it looks at about the same level as I have a sensor on my garage door at the lower hinge for when the car is coming in and out, and the raccoons are eating it. So I think of the brainstorm of putting it on the upper hinge, okay? Now I just think that how could I get a patent for that, and that— now that's very naïve, that's very naïve...Mr. Asano*

himself, I would think at some point when the Ford Company decides to switch to electronic throttles, of course will have every motivation in the world to do precisely what [Engelgau] did, because he can't use that thing that pulls back and forth anymore. Rather, he has to get a little electronic cap and attach it to something that moves in fixed proportion to the accelerator going up and down. Now those are my whole reactions when I saw this and I began to think it looks pretty obvious.

This statement in the oral argument didn't leave much doubt about where Justice Breyer stood in deciding this particular case.

10.4.5 Claim 4 collapses to an invalid state

The Court in *KSR* reversed the CAFC's ruling and found claim 4 to be obvious based on the prior art of Asano and on the fact that, in view of the copious related teachings in the prior art, one skilled in the art would naturally think to combine Asano with a pivot-mounted electronic sensor to arrive at the claimed invention. It wasn't even a close call; all the Justices decided in favor of KSR.

In its decision, the Court rejected the narrow use of the TSM test, saying in essence that it was merely one of many different inquiries that could be used under the *Graham* factors analysis. The Court held that it was not to be applied as the exclusive test because it was inconsistent with the expansive and flexible approach required for analyzing obviousness. The Court also reiterated its holdings in its previous obviousness-related cases. The following excerpts are taken directly from the KSR opinion:

- *The combination of familiar elements according to known methods is likely to be obvious when it does no more than yield predictable results.*

- *If a person of ordinary skill can implement a predictable variation, [then obviousness] likely bars its patentability.*

- *If a technique has been used to improve one device, and a person of ordinary skill in the art would recognize that it would improve similar devices in the same way, using the technique is obvious unless its actual application is beyond his or her skill.*

- *A court must ask whether the improvement is more than the predictable use of prior art elements according to their established functions.*

- *Rejections on obviousness grounds cannot be sustained by mere conclusory statements; instead, there must be some articulated reasoning with some rational underpinning to support the legal conclusion of obviousness.*

- *The obviousness analysis cannot be confined by a formalistic conception of the words teaching, suggestion, and motivation, or by overemphasis on the importance of published articles and the explicit content of issued patents. The diversity of inventive pursuits and of modern technology counsels against limiting the analysis in this way. In many fields it may be that there is little discussion of obvious techniques or combinations, and it often may be the case that market demand, rather than scientific literature, will drive design trends.*

- *Granting patent protection to advances that would occur in the ordinary course without real innovation retards progress and may, in the case of patents combining previously known elements, deprive prior inventions of their value or utility.*

10.5 USPTO EXAMINATION GUIDELINES FOR OBVIOUSNESS

In response to *KSR*, the USPTO issued examination guidelines for patent examiners to follow when performing an obviousness analysis. The guidelines provide seven rationales for determining obviousness of a claimed invention in light of the prior art:

USPTO Seven Rationales for Determining Obviousness

1. Combining prior art elements according to known methods to yield predictable results
2. Simple substitution of one known element for another to obtain predictable results
3. Use of a known technique to improve similar devices (methods or products) in the same way
4. Applying a known technique to a known device (method or product) ready for improvement to yield predictable results
5. "Obvious to try"—choosing from a finite number of identified, predictable solutions, with a reasonable expectation of success
6. Known work in one field of endeavor may prompt variations of it for use in either the same field or a different one based on design incentives or other market forces if the variations would have been predictable to one of ordinary skill in the art
7. Some teaching, suggestion, or motivation in the prior art that would have led one of ordinary skill to modify the prior art reference or to combine prior art teachings to arrive at the claimed invention

---

For a patent examiner to support an obviousness rejection, he or she need cite only one of the rationales. However, this rationale must be clearly articulated. An obviousness rejection that simply states that the invention is obvious in light of the prior art is considered conclusory and fails to provide a suitable rationale.

If a patent examiner can make an obviousness rejection that is supported by the *Graham* factual findings and the appropriate rationale, then he or she is considered to have established a ***prima facie*** case for obviousness. This means that the patent office has met its threshold burden of proof to show that the invention is obvious. At this point, the burden of demonstrating that the invention is non-obvious shifts to the applicant. This usually involves the applicant demonstrating that the rationale was improperly established or citing overriding secondary considerations, which we discuss in greater detail next.

The USPTO's obviousness guidelines also have a number of important caveats when it comes to combining and/or modifying prior-art references to arrive at a claimed invention in an effort to declare it obvious. Some of these caveats are

1. The proposed modification cannot render the prior art unsatisfactory for its intended purpose.

2. When trying to arrive at the claimed invention, one cannot combine references that together include all the claim limitations but that when the prior art inventions are combined do not operate in the manner of the claimed invention.

3. The proposed modification cannot change the principle of operation of a reference.

4. There must be some reasonable expectation of success in combining the elements.

5. The person of ordinary skill in the art is a hypothetical person who is presumed to have knowledge of all the relevant art at the time of the invention. Factors that may be considered in determining the level of ordinary skill in the art can include:
   a. The type of problems encountered in the art
   b. The prior art solutions to those problems
   c. The rapidity with which innovations are made
   d. The sophistication of the technology
   e. The educational level of active workers in the field

Sections 2141 through 2146 of the MPEP go into exquisite detail of how examiners are supposed to formulate obviousness rejections and all of the different considerations that apply. A searchable version of the MPEP is available online at http://www.uspto.gov/web/offices/pac/mpep/. If you really want to get to know all the ins and outs of what obviousness is all about, this is probably the best reference to read.

10.6 OBVIOUSNESS REJECTIONS

At this point, you might have the impression that the examination guidelines for obviousness and the Supreme Court's relatively recent clarification of the obviousness standard make it easy for the USPTO to determine whether a given claim should be deemed obvious. Were this to be true, however, patents subject to litigation would always be upheld as being nonobvious, which is certainly not the case. We can assume, therefore, that the USPTO's obviousness filter remains inefficient despite the recent issuance of the USPTO guidelines and the Supreme Court's clarifying opinion on the matter.

What does all this mean for a technology business that is trying to create a patent portfolio around its products? It means that one cannot simply rely on a classical patent mechanics view of the patenting system and assume that the obviousness standard will be applied uniformly and fairly to all patent applications passing through the USPTO. The uncertainty inherent in the obviousness filter means that some patents will issue that, in a classical patent mechanics world, would never issue. So a technology business needs to be careful about prejudging the obviousness of its own innovations when deciding to file or not to file patent applications.

This does not mean that a technology business needs to file patents on all of its innovations, including the most trivial. Technology businesses that do so usually either have no sense of how to leverage IP properly and work to a budget or have a patenting strategy that mandates that they patent everything. But it does mean that if an innovation has the potential to be very valuable to the business, that value might outweigh someone's opinion about the innovation being "obvious." This is where the balance between business, legal, and technical viewpoints about innovations needs to be struck. The very fact that an innovation has potentially significant business value may give rise to important "secondary considerations," which are discussed in detail later. But perhaps more importantly, it allows for the always nonzero, patent quantum mechanics possibility that the patent office will find it nonobvious and therefore patentable.

Obviousness rejections can sometimes be difficult for a patent applicant to deal with because, as they say, hindsight is 20:20. Many innovations tend to look obvious after you explain them. For example, we saw before that a combination of known elements that yields an "unexpected result" is a strong indicator that the invention is nonobvious. However, "unexpected results" can be hard to appreciate in hindsight. After all, if you didn't expect the result you obtained, why would you have combined the elements in the way you did in the first place? Unless you did so by pure accident (which can certainly happen; the inventions of vulcanized rubber and corn flakes are two prime examples), you must have been trying to get the result or something close to it. And why exactly wasn't it obvious to try? If the patent application doesn't explain this, then the patent applicant is left trying to make arguments that aren't directly supported in the patent application.

Some patent examiners fall into the trap of using the claim being examined as a template to piece together parts of claims from different prior-art references, some or all of which are tenuously related, if not entirely unrelated, to the invention at hand. Consider by way of example the zone 4 claim of my now famous laser toaster of Chapter 7. Recall that the zone 4 claim included features such as light-emitting diodes (LEDs) or laser diodes, germanium-oxide lenses, and an optical fiber toasting sensor.

Now, arrays of LEDs can be found in, say, some lighting fixtures. Laser diodes can be found in patents involving compact disc players. Germanium-oxide lenses can be found in patents related to infrared cameras. Optical-fiber sensors are disclosed in patents involving fiber-based gyroscopes.

You get the picture. Without reading and understanding the claim *in its entirety,* it becomes just a list of parts that can be identified in disparate prior-art references. Treating a claim as if it describes a box full of unconnected parts is not permitted. This is a classic ***claim-template reconstruction*** of a claim and is not allowed under the obviousness guidelines. Yet, it occurs fairly frequently in claim rejections. Too many claim-template reconstruction claim rejections and hindsight rejections are why the TSM test held sway for such a long time.

On the flip side, patent applicants tend to get confused and forget that obviousness is directed to the ***claimed invention*** and not to the invention as described in the application. It is important to remember the following key point about patents with respect to what is described versus what is claimed:

> **An invention can be inherently patentable but claimed in a manner that renders it unpatentable.**

It is also worth reemphasizing the point we made in Chapter 7 that zone 1 no-shame claims and zone 2 offensive claims find their way into patent applications when the patent owner (or, more likely, his or her attorney) seeks to capture all the IP space he or she can, right up to the very limits of the prior art and sometimes beyond—which is to say, to the limits of patentability and sometimes beyond—regardless of how the patent is ultimately intended to be leveraged. When those involved in the patent application process do not know the full scope and content of all the prior art (which is often the case), or when they purposely seek more property than that to which they are entitled, they are likely not only to capture more potential infringers but also to wander into the sphere of obviousness.

10.7 THE "SECONDARY CONSIDERATIONS" WORMHOLE

The Supreme Court in the *Graham* case, which set forth the ever important three *Graham* factors described before, also enunciated "secondary considerations" to be weighed along with the *Graham* factors when assessing obviousness. Secondary considerations can include:

- Evidence of commercial success
- Industry praise/acclaim
- Unexpected results
- Fulfillment of a long-felt need
- Failure of others to solve the problem the invention solves
- Whether others have copied the invention
- Industry skepticism

The CAFC has since held that secondary considerations are not limited to those cases in which patentability is a close call. It has observed that secondary considerations are not secondary in importance and that, to the contrary, they are often the most probative objective evidence available for assessing obviousness.

Secondary considerations are an important aspect of the obviousness analysis in that, when backed up by the right evidence, they can create a wormhole in IP space. The secondary considerations wormhole can transport an invention from inside to outside the obviousness sphere with the surprising effect of making an otherwise unpatentable invention patentable.

Of the secondary considerations, commercial success is perhaps the most interesting in that an invention that starts out as obvious can morph into a nonobvious invention while the patent examination process is ongoing. This is because it typically takes years to get a patent issued, and during that time a product embodying the claimed invention may be designed, manufactured, marketed, and sold in huge numbers.

One caveat is that the secondary considerations must establish a connection between the patented invention and the particular consideration being asserted. For instance, citing commercial success as a secondary consideration requires showing that the commercial success was due to the claimed invention itself and not some other reason. For example, if the patented product is part of a larger machine, it may be that the commercial success is owing to that larger machine. Or, commercial success may be attributed to an advertising blitz or other artificial circumstances that are unrelated to the product covered by the claim at issue.

Let's imagine that a technology business has an invention it feels could be very important but also thinks is obvious and would get rejected by the patent office. It chooses **not** to file a patent application on the invention. The invention ultimately becomes wildly successful and is used by other companies to solve a problem that others had tried but failed to solve. While the business might benefit from the sale of its invention, it won't be able to stop others from making and selling the same invention.

Now let's imagine that the technology business took the chance of filing a patent application on the invention despite its concerns about its being obvious. The application sits in the patent office for 3 years, during which time the invention becomes a commercial success by virtue of its wonderful attributes. Then, the first office action arrives that rejects the claims to the invention for obviousness. At this point, the business can cite secondary considerations relating to commercial success to make a case that the invention is in fact not obvious. If the technology business can get the patent, it is by definition going to be a valuable one because it can be used to protect that very same commercial success that others will certainly want to get a piece of.

10.8 WORMHOLE LIMITATIONS

At this point, it may appear that you can get around an obviousness rejection simply by establishing sufficient facts (e.g., commercial success, fulfillment of a long-felt need, etc.) to create a secondary considerations wormhole to escape to a nonobvious region of the IP space. If only it were that easy. To see why things aren't that simple, we need to take a brief detour back through quantum patent mechanics.

Under a classical patent mechanics view, we tend to think of obviousness as a binary concept: Either something is obvious or it isn't. The plot of Figure 10.3(a) graphs obviousness $O(I)$ as a function of "inventiveness" from the viewpoint of classical patent mechanics. An invention exists in one of only two states: obvious or nonobvious.

The invention switches abruptly from the "O" state to the "N-O" state at some switchover point as the degree of inventiveness proceeds from 0 (i.e., so trivial the

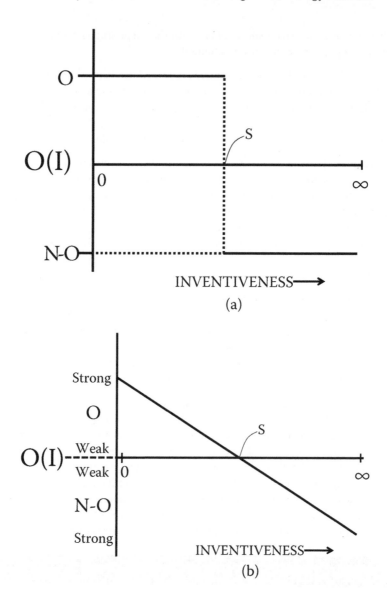

Figure 10.3 (a) Classical patent mechanics plot of obviousness versus inventiveness. (b) Quantum patent mechanics plot of obviousness versus inventiveness.

inventor should be hit smartly with a stick) to infinite (i.e., so profound as to defy human comprehension). By this classical patent mechanics measure, a prima facie case for obviousness has a uniform strength (say, normalized to 1) as soon as it can be established, and the degree of inventiveness (or lack thereof) does not matter once the switchover point is crossed. The switchover point S represents a kind of obviousness singularity that remains unexplained by classical patent mechanics.

The quantum patent mechanics viewpoint permits a more realistic view of how obviousness varies with inventiveness. Figure 10.3(b) is the quantum patent mechanics version of the same obviousness versus inventiveness plot of Figure 10.3(a).

Under the quantum patent mechanics view, there are still only two states—O or N-O, but there are also continuous degrees of obviousness and nonobviousness within each of these states. There is a smooth transition between the O and N-O states at the switchover point S, which represents the case where the claim is both O and N-O at the same time. Under a quantum patent mechanics view, the degree of inventiveness defines a relative strength of obviousness or nonobviousness, with little inventiveness relating to "strongly obvious" and extreme inventiveness relating to "strongly nonobvious." This means that the examiner's prima facie case for obviousness can range from very weak to very strong, depending on the amount of inventiveness expressed by the claim being examined.

Despite what the CAFC and the Supreme Court have said about secondary considerations in the past, it is hard to form a secondary considerations wormhole when there exists a strong prima facie case for obviousness. Arguing nonobviousness based on secondary considerations involves the filing of evidence and affidavits beyond the more familiar activity of scrutinizing other patents or prior-art publications. It can be difficult for a patent examiner to make sense out of the evidence and apply it to the claimed invention even with the MPEP examination guidelines at hand. Perhaps, most importantly, the secondary considerations analysis first requires the examiner to be able to recognize how strong the prima facie case really is, and the likelihood of an examiner admitting that it is not all that strong is rather small.

This confluence of factors makes arguing against obviousness based on secondary considerations a challenge. A journey through the secondary-considerations wormhole is likely going to include a detour through the Board of Patent Appeals and Interferences and maybe federal court if the invention is to make it all the way to the promised land of nonobviousness.

That said, when an invention proves to be extremely valuable, it may be worth trying to make the trip. In a fairly recent CAFC case involving a patent owned by Transocean Offshore Deepwater Drilling, Inc.,[4] secondary considerations overcame a finding of invalidity based on obviousness in the district court. The fact that Transocean obtained a $15M judgment meant that the patented invention had substantial value. Note also that the case had to go all the way to the CAFC to get good measurement of the patent wavefunction.

10.9 INVENTION QUENCHING

In a technology business, who gets to decide whether a given invention is too obvious to file as a patent application? As we've seen, figuring out whether an invention is truly obvious is not always easy. Even experienced patent attorneys with a good sense of the case law regarding obviousness and who know the MPEP inside and out are often reluctant to come right out and say that a client's invention is obvious and not worth patenting.

In Chapter 9, we discussed the patenting best practice of ensuring that all innovations are documented and submitted to the company for review. Innovators should not be judging the obviousness of their innovations, and there should be no prefiltering of innovations at the innovator level. This makes an innovator's job much easier because he or she is not left to wonder whether a given innovation is worthy of submitting. The answer is always "Yes, document and submit the innovation, even if you as the innovator think it is obvious."

Let's take a closer look at why this is important in the context of obviousness. Recall that analyzing obviousness requires taking into account the **level of ordinary skill in the art.** However, many innovators are people with **extraordinary skill in the art,** which means that their standard for judging whether an innovation is obvious is probably going to be too high.

In addition, as we have discussed previously, the obviousness analysis is ostensibly a legal one. While a spherical patent assumption based on logic and common sense might lead one to conclude that the lay and legal definitions of what constitutes an obvious invention are one and the same, they simply are not. The legal analysis of obviousness has developed over hundreds of years and is far more circumscribed. So when an innovator who is not trained in legal analysis decides that his or her innovation is obvious, he or she is applying a lay definition that in some cases might be right and in some cases might be wrong.

I refer to the phenomenon whereby innovators conclude that their own inventions or innovations are obvious and hence do not submit them to the company as **invention quenching** or **innovation quenching.** It is an insidious phenomenon for a technology business because the business ends up never knowing about what could be extremely valuable innovations. This situation can be agonizing when the company finds out about the innovation after it is too late, such as after it has been included in a commercialized product for more than 1 year.

10.10 THE OBVIOUSNESS/ NONOBVIOUSNESS INTERFACE

Another reason to prevent innovators from prefiltering their innovations is that some of the most valuable innovations exist at the hairy edge of being obvious and typically appear very obvious in hindsight.

A classic and very simple invention that proved enormously valuable was US Patent No. 6,281, issued on April 10, 1849, to inventor Walter Hunt of New York, NY. The figures in the patent are shown in Figure 10.4. The invention is the ever popular safety pin, which is made from a single piece of wire coiled into a spring at one end and formed as a separate clasp and point at the other end, allowing the point of the wire to be forced by the spring into the clasp.

Another simple yet amazingly important invention is the paper clip, first invented by Samuel B. Fay in 1867. His patent was entitled "ticket fastener" and his invention was simply a twisted piece of wire. Improvement patents followed, including US Patent No. 675,761, issued to inventor Johan Vaaler on January 2, 1901, which introduced the looped configuration that resembles today's oval-shaped clips. Figure 10.5 shows this well-known looped configuration.

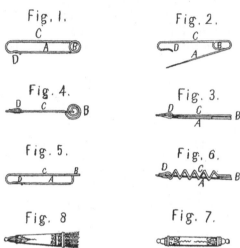

Figure 10.4 Front page of Hunt's patent on the safety pin.

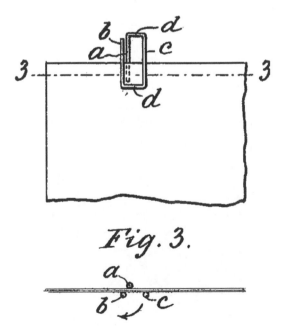

Figure 10.5 Figures from Johan Vaaler's paper clip patent.

A more contemporary example is the paper-cup holder used for holding thin cups that carry hot liquid. The invention was patented in 1995 by Jay Sorensen as US Patent No. 5,425,497. Figure 10.6 presents figures from the Sorensen patent. The cup holder (more colloquially called a "coffee sleeve") is formed as a sheet with interlocking ends that allow the holder to be wrapped around a cup.

The key feature of this patented invention is the "approximately semi-spherically shaped depressions" that each define "a non-contacting region of the band creating an air gap between the band and the cup, thereby reducing the rate of heat transfer through the holder." This is what differentiates the cup holder from a napkin simply wrapped around the cup or from coffee sleeves that have linear grooves.

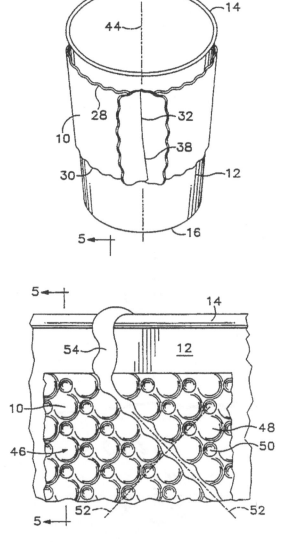

Figure 10.6 Figures from Jay Sorensen's cup-holder patent.

So here are two key points about obviousness relating to the complexity of an invention:

> Simple ≠ obvious
>
> Complex ≠ nonobvious

Many great inventions are simple and appear obvious in hindsight. Most valuable inventions aren't Rube Goldberg contraptions. Good innovators know this and do not quench their "simple" innovations; in fact, they actively pursue such innovations. A smart technology business knows this as well and thinks long and hard before deciding not to patent any innovations that it thinks might be "obvious" merely because they are "simple."

10.11 FITTING PATENTS INTO A DENSE IP SPACE

Chapter 8 introduced the concept of the fractal nature of innovation to explain how even in a densely packed IP space there is always room for innovation. However, as the IP space increases in density, it gets increasingly difficult to squeeze patents into the smaller spaces because the large amount of prior art acts as a repulsive force.

The difference between trying to insert a patent into an uncrowded IP space and trying to insert it into a dense IP space is tantamount to the difference between trying to embed a BB in a block of Styrofoam™ and trying to embed it in a block of wood. Because the wood is denser, embedding the BB takes much more energy. Likewise, it takes a lot more energy to embed a patent into a dense IP space than into a relatively rarified one.

The kind of the energy we are talking about when it comes to patents manifests in the form of high-frequency details. Two huge and related mistakes that many would-be patentees make in trying to patent their inventions in a crowded IP space are:

1. Underestimating the strength of the repulsive force that the prior art can exert against their claims; and
2. Underestimating the amount of high-frequency energy in the form of details that is required to actually embed a patent into a dense IP space.

Many would-be patentees behave as if their inventions are so pioneering as to require only the lowest frequency of disclosure (i.e., one light on details) and the broadest of claims. In spherical patent terms, broad claims and a lack of details make the spherical patent too big and slow to penetrate the available IP space.

Large, low-energy, spherical patent applications have little hope of finding their way into the tiny interstices of a densely packed IP space. Details are needed to shrink the patent application and give it the energy to make it fit and then stick. A lack of details leaves the applicant with no avenue for explaining to the patent examiner (or a court) why the invention is not obvious in view of the tightly packed prior art. Put differently, a big, broad patent application with a generalized description cannot be shrunk down because it provides insufficient information for adding limitations to the claims to reduce its size. Sometimes it is hard to know beforehand what form a rejection will take and exactly what references will be cited, and one is best positioned when a host of fallback options is available.

It takes a lot of work to flesh out the high-frequency details of an invention to give it enough energy and get it to the right size to fit onto a dense IP space. This process is asymptotic. It is easy to get about halfway to a good description of an invention because the first part of the description can be thought of as based on low-frequency generalities. This amount of effort gets you the large, low-energy patent application. Getting to a really great description that yields a compact, high-energy application requires high-frequency thinking, which in turn requires serious effort.

Gathering the high-frequency details a patent application needs to distinguish it from the prior art may require actually building the invention and conducting experiments that measure its performance. It may require explaining at some length how the combination of known parts generates an *unexpected* result, with the aforementioned measurements supporting that result. It may mean taking the time to compare the invention to prior art inventions to point out the shortcomings of the previous approaches. It may mean explaining why the problem being addressed has yet to be solved and why other approaches have failed.

This is one of the reasons why it is usually more difficult to obtain a patent on an improvement invention than on a pioneering invention, which can be described and claimed more generally.

NOTES

1. See *Hotchkiss v. Greenwood*, 52 US.(11 How.) 248 (1851).
2. See *Great Atlantic & Pacific Tea Co. v. Supermarket Equipment Corp.*, 340 US 147, 152 (1950).
3. See *Sakraida v. Ag Pro, Inc.*, 425 US 273, 282 (1976).
4. *Transocean Offshore Deepwater Drilling, Inc. v. Maersk Contractors USA, Inc.*, Case No. 11-1555 (Fed. Cir., Nov. 15, 2012) (Moore, J.).

CHAPTER 11

INVENTIONS AND INVENTORS

11.1 WHERE AND HOW DO MOST TECHNOLOGY WORKERS LEARN ABOUT IP?

Here is a simple question with profound implications: Where and how do most technology workers learn about IP?

Here is the answer for the vast majority: on the job and on the fly.

Precious few technology workers actually arrive at their place of employment prepared to work with IP—patents in particular. This situation is unfortunate, but it's not surprising. There are very few substantive college courses offered on IP. Most graduate-level engineering schools offer no IP-related classes. Only a small fraction of business schools offer a serious IP course.

Somewhat ironically, some of the best formal IP training is provided by law schools. Some law schools offer a master of laws degree (called an LLM) in IP. This goes a long way in explaining why the world of IP has lots of lawyers in it. So, while there are certainly some notable exceptions, the teaching of IP in academia is generally wanting in the places where it is needed most: engineering, science, and business departments at academic institutions that are training future inventors and innovators.

11.2 THE IP ELEPHANT IN THE ROOM

When technology workers learn about IP on the fly, much of their knowledge is based on IP folklore, which, like other kinds of folklore, gets distorted and embellished as it is passed down from generation to generation of technology workers. Some of the folklore is pernicious and takes the form of extreme IP negativity. I have heard people with absolutely no IP experience or knowledge declare, "Patents are worthless" and "Patents don't matter because another company will just steal your invention if they really want it." This is not the kind of thinking a technology business wants to be promoting in its workforce if it is trying to get patents to protect its products or for licensing purposes.

The ad hoc, on-the-job approach to learning IP is outdated and unsuitable for a modern technology business. The tuition at the School of Hard Knocks is too high, and the consequences of screwing things up are too great. It is a Flintstones approach to trying to get technology workers involved in a Jetson's IP world.

So now, the IP elephant in the room: The lack of serious IP education opportunities for technology workers has led to a high rate of IP illiteracy in the technology workplace. This illiteracy goes a long way toward explaining why so many technology businesses have high-entropy and manifestly dysfunctional patenting systems. How can technology workers be expected to participate effectively in the innovation process when they have inadequate knowledge of the underlying subject matter that the process is designed to handle?

Based on my dozen years or so of talking with many hundreds of inventors at a range of technology businesses (including academic institutions), I would estimate that only about 10% of the technology workforce is IP literate. By IP literate I mean having a working knowledge of IP to the level needed to carry out their job responsibilities properly as they relate to IP matters, to participate effectively in their organization's patenting system, to interact efficiently with patent attorneys, and generally to operate in a manner that decreases rather than increases patenting system entropy. While this number will certainly vary among technology businesses, it is the rare business whose workforce has an IP literacy rate approaching anywhere near 50%.

If a technology business just assumes that its workers will pick up what they need to know about IP on the fly, then that's exactly what they'll do. The workers' knowledge will reflect the incomplete nature of that kind of learning process. A few people that care enough to learn about IP may indeed become quite knowledgeable on their own steam. They will likely constitute the small group of inventors that do most of the inventing at the company. Sound familiar?

To be sure, some technology businesses offer solid IP training. For example, when I first began my job as an engineer at IBM, I had no clue about IP anything. Fortunately, I met people who really knew something about patents, and they were a tremendous resource for me. IBM's impressive internal education system also provided me with opportunities to attend classes and workshops. IBM's management was also fairly proactive in not only promoting the availability of the IP education opportunities but also insisting that they be attended for professional development.

IP Education at Boston University

Dan Cole is an associate professor of mechanical engineering at Boston University. Prior to joining the faculty at Boston University, Dan worked at IBM. Based on his experience with intellectual property matters from his years of working in industry, Dan decided to start teaching an IP course at Boston University called "Invention: Technology Creation, Protection, and Commercialization." The course description can be found on Boston University's website at http://www.bu.edu/me/me-502-intellectual-assets/. The course covers such things as patent searching, patent analysis, the legal aspects of patents, and business models for commercializing intellectual assets.

Dan's course is one of the few graduate academic courses offered by an engineering department and directed to teaching practical IP subject matter to would-be technology workers. The class averages about fifty students, divided roughly between graduate students and seniors. Dan's course can serve as a model for other academic institutions that have an interest in seeing their engineering and science students be IP literate when they enter the technology workplace.

11.3 IP TRAINING AND EDUCATION

Technology businesses need to consider providing their employees with the IP tools not to just survive but also to thrive in the IP universe. While not all companies have the vast financial resources and educational infrastructure of an IBM, no company can reasonably expect to have an efficient and effective IP system without providing its technology workers with substantive IP education. This includes small companies and start-up companies especially, since many of these kinds of companies live or die by their IP.

Some technology businesses think that hosting the occasional talk by a patent attorney is all that is needed to bring everyone up to speed. While these talks can be really helpful, they can also be really awful. Some of the worst talks about IP come care of patent attorneys who immediately start in on the high-frequency details about patent law. I can say that because I've given such talks in the past, to my regret. A bad IP presentation can turn people off and leave the wrong impression with those you want to influence most.

While an IP education and training effort could take many forms (e.g., classes, workshops, study groups, etc.), it needs to be dynamic and engaging. This means the message must be delivered through people who are dynamic and engaging. In other words, keep the boring people away from the microphone. IP is too important for a business to risk botching the delivery of the message to the key people in any IP effort—the technology workers. A half-hearted effort at IP education will only reinforce the old and unfair stereotypes of patents as boring and the exclusive purview of asocial and introverted patent attorneys. The IP Era of the Information Age requires an updated and relevant message.

In my view, the best IP training comprises three main sections. The first section introduces the low-frequency, big-picture view of what IP is and how businesses use it. An IP-savvy businessperson is the best person to tell this story. People need to hear about patent licensing and the revenue it generates. They need to hear about a specific case of patent litigation and how the process and the outcome impacted the business. They need to hear about what innovations proved to be the most valuable to the company, who generated them, and what led to the innovations. They need to hear about what their competitors are patenting and how these patents are affecting their own company. They need to hear interesting stories about their company as they relate to its patenting efforts, and these stories need to tie in with the company's bottom line. In short, technology workers need to hear good motivational speakers who can explain how patents are leveraged as business tools and how that leverage benefits the business.

In the training sessions I have been asked to organize, I try to get the CEO to serve as the first speaker. If the CEO is not available, then I get the CTO. Every CEO and CTO that I have heard speak on IP has done a great job; they shared stories about the IP big picture that ordinary technology workers don't usually hear or think about. CEOs and CTOs tend to be truly excited and passionate about the importance of IP in their business and want everyone to participate with enthusiasm. Their enthusiasm tends to be contagious.

The second section of the IP training delivers the midfrequency, operational picture of the business's IP management and IP process. Ideally, this information is presented by, say, the IP project manager in charge of driving the patenting process for the business. This information fills in the picture of how the IP system is supposed to operate, who is in charge, where to get access to forms, how to find an IP mentor, how to prepare a proper innovation disclosure, and so on—all the nuts and bolts of the best IP processes and practices. It explains how technology workers fit in and how they can participate. Too many technology workers end up as bewildered bystanders to their business's patenting system because they don't know the practical aspects of how the patenting system works. Nothing about it is complicated, but participating does take some organization, discipline, and ability to follow established procedures. And in order to be followed, these procedures need to be explained effectively.

The third section of the process involves the patent attorney. Now is the appropriate time to invoke some of the high-frequency, legalistic information about patents that would-be innovators need to know. The first two sections provided the proper high-level perspective that should motivate technology workers to get into and understand some of the detailed legal underpinnings of patenting. While the legal-based material they need to cover is detail intensive and may be not all that fun to listen to, becoming truly IP literate requires some pain and attention to the particulars.

11.4 IP AND TECHNOLOGY WORKER PROFESSIONALISM

Many technology workers operate under the misconception that innovating and inventing are supplemental to their normal work and fall outside their basic job description. This misconception arises because of a fundamental misapprehension of the nature of a technology job in the IP Era of the Information Age. This misconception is often reinforced by the way the technology business manages its patenting system and the expectations it sets for its technology workers.

It is hard to say exactly when the IP Era of the Information Age began, but what is clear is that it is here now. We are in it. Technology workers who believe that inventing and innovating are "extra work" and peripheral to their central tasks are hugely disconnected from the reality of working in today's technological society. Innovating, inventing, and participating in a technology business's patenting system are now integral parts of a technology job in today's economy. What might fairly have been considered extra work during the Industrial Revolution is now a natural and expected part of the job in the technology revolution. To the extent that the "extra work" mind-set prevails and is consciously or subconsciously promoted within a technology business, that technology business will suffer the consequences of having an Industrial Revolution mind-set about IP.

11.5 IP ZANSHIN

Zanshin (残心) is a Japanese word that in martial arts speaks to the concept of having such total awareness that you can act without thinking. Zanshin is used to refer to an empty state of mind where taking action requires no thought. In a confrontation, there is no time to stop and think about all the steps required to perform a movement or achieve a position. In Brazilian jiu-jitsu, for example, there is a saying that goes something like this: "If you have to think, you're already late. If you're late, you need to use muscle. If you need to use muscle, you tire. If you tire, you lose." The ability to act without having to process information consciously is a goal of many martial arts and signifies a true martial-arts master.

The concept of zanshin has applications beyond martial arts. The average person has likely developed a high degree of zanshin in a number of common activities. A simple example is typing into a computer; at some point, you stop thinking about the actual letters and numbers on the computer keyboard. The words flow from your mind directly to your fingers, which move almost at the speed of thought. To type at such speed, there is no time to contemplate the identity and location of every key and its associated finger. In achieving a state of typing zanshin, the keys disappear and moving hands magically make text appear on the computer screen. The extra step of having to think about the actual keys is eliminated.

In a more refined example, good musicians also practice zanshin. They go beyond thinking about which positions on their instruments are associated with which notes and instead open a direct channel between their minds and their instruments to create beautiful music.

A worthy goal of a technology business is to help its technology workers achieve a state of *IP zanshin*. A technology business can greatly benefit from establishing a culture where technology workers are educated, trained, and exposed to IP until it becomes second nature. Those who achieve IP zanshin no longer need to think about the actual steps involved in innovating and participating in the patenting system in a manner that follows IP best practices. When confronted with technical problems in their work, they already know that the solution may involve innovations that could lead to a patent or trade secret. They no longer wonder whether to write up an innovation disclosure; they simply write it and submit it. They review drafts of patent applications as a matter of routine and get right to the main issues that need to be addressed without complaining about the drudgery of it all. They know how to work with co-inventors in a coordinated manner. They execute patent application documents correctly the first time and return them without delay or excuse.

From an IP-zanshin perspective, the patenting process is not separate from the technology worker's job; it is one with the job. The patenting process, rather than being filled with angst and overanalysis, is imbued with a Zen-like calm and focus. It just flows because there is a deep understanding and respect for what the

IP represents and how it needs to be handled. This level of IP knowledge gives the technology worker confidence in dealing with IP issues as they arise. The patenting process has a sense of harmony (which is the Eastern way of saying that the patenting system has low entropy) rather than a sense of discord.

11.6 AWARDS AND REWARDS

Many technology businesses seek to promote participation in their patenting systems by offering their technology workers cash rewards. A typical invention reward program pays each inventor a sum of money when a patent application is filed based on the inventor's innovation disclosure. Some programs offer an additional reward if a patent issues. Some offer rewards if the review committee decides to keep the innovation as a trade secret. Amounts typically range from fifty to several thousand dollars. Inventors at some companies make tens of thousands of dollars a year in extra income from the company's invention rewards program.

Would-be inventors tend to like an invention rewards program for obvious reasons—extra money for what is often perceived as extra work. Business managers like these programs because they feel they motivate their employees to do something that the managers normally have trouble getting them to do—namely, to participate in the patenting system.

Invention rewards programs can add entropy to a patenting system when the unintended message it sends is that submitting innovations to the company is an optional and extra-credit activity. After all, a reward or bonus is universally understood as representing performance above and beyond expectations. The message that many innovation rewards programs send is: If you don't want the extra money, then don't submit your innovations.

The problem here is that many of today's technology workers will actually opt not to participate. A fair number of the technology workers I know purposely forgo work tied to their bonuses because they would simply rather not have to work nights and weekends for the extra money. They purposely opt out of the bonus program and skip the money because to them the free time is actually more valuable.

Rewards systems also tend to assume a spherical employee. They treat all technology workers in the same way—whether a scientist working on solving cutting-edge problems and inventing every day or a technician running tests on a manufacturing line or a manager swamped in personnel management and customer and vendor issues. A rewards system can end up being a special program for a select group in the company that are best positioned to leverage the benefits of the system. Those who feel that they won't ultimately benefit from the rewards program because they can't play the game as well as the others will tend just to sit the game out.

11.6.1 Case Study: The Employee Inventors Law in Germany

There is a law in Germany called the Employee Inventors Law (EIL) that covers situations relating to employee inventions. The EIL states that inventions developed by employees of a company are owned by the employee. However, the EIL requires employee-inventors to disclose their inventions punctually and completely to their employer by submitting an invention report. The employer then has the right to claim the invention in exchange for monetary compensation to the employee. The employer must make the claim within 4 months of receipt of the invention report or the employee will retain ownership of the invention.

If the company chooses to claim the invention, the employee's compensation is royalty based per guidelines established by the German courts. Furthermore, the company cannot just sit on the invention and avoid paying compensation. Once it claims an invention, the company becomes obligated to pursue patent protection. Disputes between an employee and an employer about an employee's invention, the amount of compensation, the patentability of the invention, and so on are handled first by an arbitration committee in the German Patent Office and only then may be addressed in court.

Unlike Germany, the United States has no statutory law that requires a US employee-inventor to disclose inventions to his or her employer. However, US law does similarly concur that in the absence of an express agreement to the contrary, employees generally own their inventions, with exceptions relating to employees who were "hired to invent." However, most of these agreements do not specifically state that the employee is compelled to disclose his or her inventions. Rather, many just state that the inventor has an obligation to assign all of his or her inventions to the company. Also, US companies have no legal obligation to share the profits the invention generates with the inventor.

11.7 THE IP PERFORMANCE METRIC

One approach to incentivizing technology workers to participate in their business's patenting system is to make the obligation to submit innovation disclosures a component of the performance metric used to establish the employees' annual increases and bonuses. The philosophy here is axiomatic but profound in its simplicity: *People will perform to the standard by which they are measured.*

An IP performance metric is based on the nature of the particular technology worker's job and level of experience. Under the IP performance metric approach, employees are no longer spherical. Rather, the different ranks of technology workers are subject to different expectations. A technician who is not being asked to solve specific problems is going to have a different metric for submitting innovation disclosures than, say, an engineer who was hired to invent new kinds of machines, assemblies, and subsystems. A junior engineer might be required to submit a few innovation disclosures a year, whereas a senior scientist with 20 years of experience and who is in the thick of having to invent might be expected to submit a dozen or more.

If the junior engineer ends up submitting six innovation disclosures, then he or she has surpassed expectations and should be rewarded in kind. If the senior scientist ends up submitting six invention disclosures, then, with all other things being equal, he or she would be considered as underperforming and *not* entitled to *any* reward or bonus.

The point here is that once the submission of innovation disclosures to the business is tied to an IP performance metric that accurately reflects a technology worker's position, the message is loud and clear: Disclosure generation and submission are nonoptional. Further, if the employee exceeds expectations and goes above and beyond the metric, then his or her compensation can be increased accordingly. The reward is given out only for exceeding the set expectations and not for doing simply what is expected.

Of course, business common sense must also apply. If a senior scientist submits three invention disclosures when twenty were expected, but one of the three turns out to have enormous value, then the metric need not be slavishly followed. A benefit of having an IP performance metric is that it can be flexible and made to fit the individual circumstances more than could, say, a straight-up, across-the-board reward system that is based only on numbers and spherical employees.

Another benefit of the performance metric relates to inventorship. It may be that only a very small number of a relatively large group of people involved in a project end up getting named as inventors. What, then, about all the other people in the group who worked hard to get the project implemented? They won't benefit from an invention rewards program. However, their contributions can be accounted for in a performance metric that is based on each individual's position and expected role.

The IP performance metric can also be combined with an invention rewards/awards program: The two approaches need not be mutually exclusive. The overlay of the IP performance metric on a rewards/awards program will at least serve to dispel the notion that participation is optional.

11.8 CHANGING THE IP CULTURE

One of the hardest things to do in a business is to change an aspect of its culture. A major benefit of having an IP performance metric is that it provides an effective mechanism for changing the IP culture of a technology business. If senior management can put together an effective IP plan and articulate the IP message, then the IP performance metric can provide an effective vehicle for communicating the IP message from the low-frequency executive level to those working at the higher frequency detail level.

This communication is optimized by evaluating the manager of each technology worker according to his or her own performance metric that measures how the manager communicates and reinforces the IP message. For example, during employee meetings or employee reviews, managers can keep their people on track by inquiring about progress on IP-related matters. When a technology worker comments, "We've made several key improvements to the system," the manager needs to follow up with something like, "That's great—so have you documented the improvements and submitted them to the IP department yet?" If the technology worker responds, "Well, I didn't fill out the standard innovation disclosure form and instead just wrote a white paper," the manager must make it clear that this is unacceptable. A manager who has heard this one too many times might rightly respond, "We have innovation disclosure forms for fundamental business reasons that you don't get to override unilaterally. Please stop the white-paper nonsense and go fill out the proper form using the proper procedures and submit it by the end of the week."

Senior managers that manage other managers also need to have an IP performance metric based on how their managers communicate and reinforce the IP message as well as on measurable IP results. It is not uncommon for a technology business to have some managers that simply don't believe in IP and don't want to buy in to the IP message. However, if the company is serious about the IP performance metric, then the performance of such managers ought to be conspicuously out of line with that of their peers.

11.9 AWARDS VERSUS REWARDS

Rather than just giving out rewards (or in lieu of rewards), the IP performance metric can serve as the basis for providing *awards* that recognize outstanding performance. The award might recognize, for example, the best invention of the year, the most creative invention of the year, the most successful invention from a licensing viewpoint, and so on. An invention award that recognizes excellence in innovating is going to be viewed differently than a cash-for-inventions program that looks more like bribery than a proper incentive.

Some technology businesses recognize inventors at an annual awards dinner. At some of those dinners, the business executives hand out relatively large cash distributions (e.g., $10K to $100K) to inventors whose patented inventions have turned out to be very valuable to the business. Some technology businesses even choose to give the inventors a percentage of the royalties from licensing the patented invention.

11.10 THE IMPORTANCE OF INVENTORSHIP[1]

As we learned earlier, one of the requirements for a patent is proper inventorship. A patent obtained in the name of a person who is not actually an inventor can be invalidated for misjoinder of inventors. Likewise, a patent that omits an actual inventor can be invalidated for nonjoinder of inventors. Given that misjoinder and nonjoinder of inventors can blow up the entire patent—not just a claim or two—it is important to get the inventorship right.

At one time or another most scientists, engineers, and technicians have been involved in work that results in a published article. Depending on the circumstances, the list of authors may or may not correspond exactly to the names of those who carried out the research. In this sense, one might observe that the concept of "authorship" is rather loosely defined. Furthermore, the lay notion of "inventorship" is not the same as its legal definition, and confusing the two can create problems.

To get a better sense of how the inventorship of an invention is determined, let's consider a fictional technology business called Gargantucom. Gargantucom is a high-tech telecommunications company specializing in fiber optics. Dr. Nicole Prism is a group leader in Gargantucom's Fiber Optics Laboratory. As we arrive on the scene, she has just called a meeting to discuss a problem involving the optical transmission of the group's latest optical-fiber cable. Dr. Prism called the meeting because she is under pressure from the lab director, Brewster Angles, to come up with a solution for getting irate customers off his back.

During the meeting, an optical engineer by the name of Mike Rolenz opines, "Wouldn't it be great if we had a way to amplify light as it traveled through the fiber?" Upon hearing this, Dr. Prism exclaims, "Excellent suggestion! Say, I have an idea. Let's try soaking the fiber in red wine. I think the alcohol and acidity might favorably interact with the glass structure to give us the result we need." She then turns to her able assistant, Ian Diffusion, and says, "Ian, can you please go over to the lab and give it a shot?"

Ian thus goes off to the lab and soaks a kilometer-long strand of optical fiber in a vat of Chianti for a week. After running some experiments, Ian finds to his amazement that light traveling through the fiber is indeed amplified! He shows the results of his tests to Dr. Prism. She glances at the results and exclaims, "Excellent! We must patent this! Quick, write up an innovation disclosure so that it can be reviewed and sent to our patent attorney, Lee Galeeze, as soon as possible!"

A few days later, Ian hurries up to Lee's office and hands her an innovation disclosure listing himself, Nicole Prism, Mike Rolenz, and Brewster Angles as the "inventors." Lee sits Ian down and has him explain how the invention came about. After telling the story, Ian asks, "So who are the actual inventors?" Lee sits back in her chair and says, as lawyers are wont to say: "It depends."

11.11 THE GENERAL RULE OF INVENTORSHIP

The general rule of inventorship is that to be listed as an inventor on a patent, a person must make a contribution to the conception of the invention as reflected in at least one of the claims. This means that inventorship, technically speaking, cannot be determined until the patent's claims have been allowed by the patent office. As we discussed earlier, the claims of a patent set forth its "metes and bounds" by describing the invention in precise if not arcane language. These claims are reviewed by the patent office, and their number and scope may (and typically do) change during the examination of the patent application. Thus, it may happen that some inventors initially listed on a patent application need to be removed if the claims to which they contributed are rejected, amended, or otherwise struck from the patent application.

Nevertheless, because a patent needs to be applied for in the name of its inventors, you have to start with your best guess based on the claims as submitted in the patent application. To see how this works in practice, let's return to our friends at Gargantucom. Who are the inventors? Let's first consider Brewster Angles, head of the laboratory. When Lee asks Ian why Brewster's name appears as an inventor on the invention disclosure, Ian responds, "Well, he is the lab director, and the unwritten rule around here is that his name goes on all the publications and patents from the departments under his supervision."

This is the easiest case to decide. The answer to the question of whether Brewster is an inventor is a resounding "no." It is clear that he had no involvement with the inventive effort that resulted in the invention. Also, simply being "the boss" doesn't cut it from a legal point of view.

What about the notion of including Brewster as a professional courtesy or management policy? While this practice might fly with technical papers and other types of publications, it is simply not appropriate for patent applications. While the patent office has fairly liberal rules for correcting inventorship, committing outright inventorship fraud can have serious negative consequences. If Brewster is a company executive, the intentional falsification of inventorship might be considered a breach of his fiduciary duty to the technology business.

How about Mike Rolenz? He was the one who suggested amplifying the light traveling through the fiber. Even Dr. Prism acknowledged that this was "an excellent suggestion." Add the fact that Mike participated in the discussion and works in the group, and there is a compelling logic to including him as an inventor.

Unfortunately for Mike, he cannot be considered an inventor either. This is because all he really said was, "Wouldn't it be great if…" Such precatory statements do not rise to the level of a contribution to the invention: It is one thing to wish for something to be true and another thing to come up with a way to make it happen. Patents are all about teaching *how* to do something. Mike's contribution provided no enablement—that is, none of the *how to* needed to accomplish the desired goal.

What about Ian Diffusion? He did all the heavy lifting by actually carrying out the experiments and analyzing the results. He also kept a laboratory notebook and wrote up a description of the invention that might form the basis of a patent application. Is Ian an inventor? Again, the answer is "no." The problem with listing Ian as an inventor is that he simply carried out the instructions of another and did not personally add to the invention. It was Dr. Prism who had the idea of soaking the fiber in red wine. Ian simply executed this idea and collected data using known techniques. The general rule here is that one who simply works under or at the direction of another to execute an idea is generally not considered an inventor. This problem arises frequently in the context of laboratory technicians.

What if Ian had altered the experiment suggested by Dr. Prism? Let's say that the initial experiments with red wine showed only scant amplification. Not being satisfied, Ian adds other wines to the red wine. After exhaustive experimentation and obligatory travel throughout the wine regions of France, he discovers that adding one part Bourgogne Aligoté (1997) to twenty parts Chianti results in a much greater amplification. In this case, Ian would have made an inventive contribution that would likely end up as a claim in a patent on the invention, and Ian would thus be considered an inventor.

Finally, what about Dr. Prism? She was the one who had the idea of soaking the fiber in red wine. Her contribution was more than just wishful thinking and had enough focus to provide others with a method of improving light transmission through an optical fiber. On the other hand, Dr. Prism didn't do any of the work, did she? She just had an idea. It was Ian who carried out the "real work" in the form of experiments, and it was he who analyzed the results.

As it turns out, the kind of idea Dr. Prism had is the stuff of inventions. The fact that she turned the work over to Ian to complete under her supervision does not matter. She conceived of the way to accomplish the amplification and would be considered the sole inventor of the new light-transmission amplification method. Had Ian contributed the idea of adding the Bourgogne Aligoté, Dr. Prism and Ian Diffusion would have been joint inventors.

Correctly identifying the inventors is important if a patent application and any patents issuing thereon are to be upheld as valid. Inventorship problems can open the door to a legal challenge of a patent's validity. So the inventorship requirement is not necessarily a no-brainer requirement to meet, and spherical patent assumptions about inventorship often quickly lead to inventorship errors.

11.12 MANAGING EXPECTATIONS ABOUT INVENTORSHIP

Misunderstanding what constitutes inventorship can lead to dashed expectations on the part of those involved in a project that yields patentable inventions. Because of the relatively narrow legal requirements for inventorship—as compared to, for example, authorship—only a few of the people working on a group project might rightfully qualify as inventors, leaving the others to wonder why they were left out. If management bases its recognition of worker effort solely on the legal definition of inventorship, then the efforts of some members of the group may go unacknowledged. Managing the expectations of technology workers in the area of inventorship is important. No one wants to wait until the invention awards ceremony where checks are being handed out to find out he or she didn't qualify as an inventor.

Ensuring that technology workers understand the nature of patent inventorship through proper IP education significantly decreases the likelihood of disputes and arguments about who should and should not be listed as an inventor.

11.13 INCLUSION RATHER THAN EXCLUSION

When it comes to writing technical papers, pride of authorship and ego considerations can serve to keep potential authors' names off the paper. To the extent that this is also true of patents and inventors, a major opportunity for a better invention might be missed. A good invention by one inventor may benefit from having others scrutinize it and improve it.

If an inventor has a problem with someone "muscling in" on his or her invention, then the invention might not achieve its full potential. Engineers, scientists, and technicians need to appreciate this point and consider inviting others to join them as inventors. When inventors take an inclusive approach, the result is usually a better patent.

Let me give you an example. I am a named inventor on US Patent No. 5,680,588, entitled "Method and system for optimizing illumination in an optical photolithography projection system." The invention relates to the manufacturing of semiconductor chips by projecting light through a mask and imaging the mask onto a semiconductor wafer with very high fidelity. The basic idea behind the invention is that optimizing the illumination for the given mask pattern will optimize the image formed on the wafer. It sounds easy, but there is a lot of science behind this kind of imaging.

My co-inventor on the patent is Alan Rosenbluth. Alan is a scientist at IBM's Watson Research Center in Yorktown Heights, New York. At the time I was at IBM, he was working in the area of cutting-edge optical photolithography technologies, including illumination. He was already an experienced inventor and currently is named on over seventy-five patents.

While I came up with a basic version of the invention, I couldn't quite figure out the best way to expand upon it. I knew that Alan, on the other hand, was light years ahead of me in his knowledge of advanced mathematics and algorithms, which was just what the invention needed. I contacted Alan and asked if he would be interested in joining me on the invention and taking it beyond what I could do with it by myself. It turns out that Alan had been thinking about the same kind of invention. He was able to expand the invention to its natural limits, and his participation made for a much better patent than I could have generated on my own as the sole inventor.

11.14 THE ORDER OF INVENTORS

If there are multiple inventors on a patent application, in what order should their names be listed? Should it be in order of contribution to the invention? Is there even any significance to being listed first on a patent? The simple though not uncontroversial answer is that, unlike the order in which the authors on a paper are listed, there is no legal significance to the order in which inventors are listed on a patent. Nevertheless, it is not uncommon for inventors to get hung up on the order of the listing. After all, many inventors also coauthor papers wherein the order of authors is significant.

One way a technology business can circumvent inventor conflict about the order in which their names are listed is to make it a policy that inventors be listed alphabetically in all cases. While this might appear to give Abby Aardvark an unfair advantage over her colleague Zoltan Zyzski, from a legal viewpoint the order is irrelevant. The alphabetical order policy could save you from having to sit in a meeting listening to your colleagues argue about who did the most and why they should be listed first. The IP-literate technology worker already knows that it doesn't matter and is not going to waste anyone's time making a fuss about it.

11.15 THE COEFFICIENT OF SHAME

One of the most important things a technology worker needs to understand about the process of innovating is that even the smallest innovation can be potentially valuable. As we discussed in Chapter 10, there is a tendency among technology workers to be too judgmental about their inventions and make unilateral decisions not to submit some of their innovations to the business. This innovation quenching happens when technology workers believe their inventions to be obvious or otherwise unworthy of consideration. Some technology workers can have too much pride to submit inventions that they believe are too simple or uninteresting.

When it comes to judging their own innovations, every technology worker has what can be thought of as a coefficient of shame C_S. This coefficient has a range that extends from 0 ("absolutely no shame") to 1 ("infinite pride"). We've all met people who have a coefficient of shame $C_S = 0$. They think every thought they have is so valuable that they have to corner every person they see to praise its merits regardless of prior art dating back to Archimedes. On the other side of the spectrum are the people who have a coefficient of shame $C_S = 1$. They talk only when they believe that what they have to say is so profound it will change the world. If they came up with an invention that cured cancer, they would deem it worthy of writing up only if it were more than 90% effective.

One manifestation of IP illiteracy in a technology business is that the coefficient of shame C_S of the technology workers is on average way too high. It usually falls somewhere between 0.7 and 1 when more often it needs to be between 0.1 and 0.5. A high coefficient of shame C_S causes inventors to filter out potentially important inventions based on their assumption that their inventions are unworthy of submitting to the business or perhaps are simply not patentable.

One source of the unreasonably high coefficient of shame C_S is the misunderstanding many workers have about what constitutes a potentially important innovation. As mentioned earlier, many people think that an important innovation has to be complex while the reality is that most important innovations are simple and almost always appear obvious in hindsight.

Another source of the unreasonably high coefficient of shame C_S is based on a misunderstanding about just how tightly packed an IP space can get. The answer is that it can get much denser than most technology workers think. An IP-literate technology worker who is also an experienced inventor will submit innovation disclosures on inventions so simple that they shock their IP-illiterate co-workers. I've seen some of those IP-illiterate co-workers heap scorn on their IP-literate colleagues for trying to patent "simple" inventions. The IP-literate technology worker knows that it is up to the review board to decide what innovations to patent and the USPTO to decide whether to allow a patent. The IP-literate technology worker also knows that IP space can tolerate a relatively high density of patents by the very nature of IP space as well as the unwitting placement of patents in the forbidden regions, as we discussed in Chapter 8.

11.16 WHEN IS AN INNOVATION PATENTABLE?

The recent shift in the United States from a first-to-invent patenting system to a first-to-file patenting system makes it more important than ever to file patent applications on inventions at the earliest possible date. This raises the question of how, as an inventor, you know when your invention is ready to be the subject of a patent application.

The short answer to this question is: much sooner than most technology workers think. Technology workers tend to presume that the invention has to be embodied in a working model or that it must perform to relatively high standards of operability and reliability, such as those associated with a commercial product.

The general rule, however, is that an invention is ready for patenting when it can be described with sufficient detail to enable a person skilled in the art of the invention to make and use the invention without undue experimentation. In other words, an invention is ready for patenting when it is **enabled.** No working model is needed. Also, while operability of the invention is an inherent part of providing an enabling description, the invention itself does not have to operate up to the standard for the commercial product one may be envisioning. This is especially true if the claims do not need to include such performance limits.

Let's take a look at the relatively new IP space that was created upon the discovery of a new material called graphene.[2] Graphene is a one-atom-thick planar sheet of carbon arranged in a honeycomb lattice. It exhibits amazing physical properties such as high electrical conductivity (as good as copper), high transparency, high thermal conductivity, and high mechanical strength. Graphene is expected to find important uses in electronic circuits (e.g., as electrodes), batteries, touch screens, and solar cells. Flakes of graphene are also useful for certain applications and are constituted by layering a number of graphene sheets. In fact, the 2010 Nobel Prize in physics was awarded to Andre Geim and Konstantin Novoselov for their work in studying graphene, including their isolation of graphene from graphite using adhesive tape. The work of Geim and Novoselov gave rise to the big bang of the graphene IP space that, as is natural, now has many players rushing in to claim a piece of the action.

Because of the tremendous promise of graphene for all sorts of applications, there are huge efforts under way to produce it in its pure form. To the extent that someone invents a method of producing graphene from graphite, it may not be critical to the patentability if the method does not produce graphene in copious amounts. All that is required is that the method satisfy the requirements for patentability, in particular the novelty and nonobviousness requirements. It may not matter if the method takes days, weeks, or even months to make a single sheet of graphene.

To the extent that an improvement invention can make a slow method of making graphene faster and therefore more in line with scaling up for the commercial production of graphene, the improvement may very well be patentable.

The claims for the improvement method, however, may require limitations that speak to the "new and improved" aspect of the invention—namely, the increase in the production capacity over the prior-art method—to distinguish it from the original (i.e., prior-art) method.

So in case you missed it, here is a key point about when an invention is ready for patenting:

> **An invention can be ready for patenting even if it works terribly according to existing or anticipated commercial standards.**[3]

Remember, the first version of anything tends not to work very well. If you wait for your invention to work to perfection or even close to it, I can almost guarantee that someone else with a lower coefficient of shame will file a patent application on his or her lousy version first. You will then have to deal with what is likely to be very close prior art, and you may be put in a position of infringing the claims of the prior-art patent on the lousy version of your invention if you want to practice your wonderfully performing version.

11.17 MENTORING

One of the best ways to help technology workers develop as innovators is through a mentoring program that connects the person with little or no innovation and patenting experience to a person having substantial experience in these areas. It gives the beginner a lifeline to the innovation and patenting process. The mentor is there to answer any questions and give his or her "mentee" the lowdown on how the patenting system actually works for the organization. There is nothing like one-on-one interaction when it comes to getting people the information they need to feel comfortable working within the patenting system.

One key area in which a mentor can help a mentee is the innovation disclosure writing process. The mentor can provide sample invention disclosure forms that show the level of detail and type of information required. Mentors can also help mentees as they are writing their innovation disclosures and can provide valuable feedback on drafts. Further, if there is a review board that needs to hear a presentation on the invention, the mentor can help the mentee prepare for this occasion as well as show up and advocate for the mentee at the review board meeting.

Mentors can end up being co-inventors on inventions because they have key insights into the inventive process and usually know how to expand and develop an invention to its natural limits. This is a good thing and speaks to the advantages of the teamwork approach to inventing discussed before. It is also good that the mentee see how this process is carried out so that she or he can apply it not only to the present invention but also to the next invention.

Mentoring is also a good activity to measure in the performance plan of senior technical workers. Most senior people find the process fun and rewarding and not a burden, especially if they can remember how lost they were when they first were trying to get their innovations through the patenting system. Such performance metrics are usually easily met and the benefits to the mentees are long-lasting. Mentoring also ensures that institutional knowledge is passed down to the new generation of technology workers being hired and nurtured.

On the flip side, it is probably best not to have the mentor provide feedback to the mentee's manager. For the mentoring relationship to succeed there needs to be a kind of mentor–mentee privilege that allows the mentee to feel free to ask potentially embarrassing questions and otherwise interact freely with the mentor without the gravitas of being scrutinized and reviewed.

NOTES

1. Sections 11.10 and 11.11 are adapted from Gortych, J. E. 2003. *Legal lens anthology: Essays on intellectual property,* 58–59. Washington, DC: Optical Society of America. Copyright Optical Society of America and used with permission from the Optical Society of America.
2. See, for example, Geim, A. K., and K. S. Novoselov. 2007. The rise of graphene. *Nature Materials* 6 (3): 183–191.
3. See, for example, MPEP § 2164; see also *Inc. v. Yieldup Int'l Corp.,* 349 F.3d 1333, 1338, 68 USPQ2d 1940, 1944 (Fed. Cir. 2003). (It is not necessary to "enable one of ordinary skill in the art to make and use a perfected, commercially viable embodiment absent a claim limitation to that effect.")

CHAPTER 12

INDEPENDENT INVENTORS[1]

12.1 INTRODUCTION

The words "independent inventor" conjure up a romantic idea of a Thomas Edison-like entrepreneur working long hours in a lab until one day the long-awaited discovery is made. The inventor then shouts, "Eureka!" and runs from the lab to the patent office with patent application in hand. Shortly thereafter, a beribboned patent covering the wonderful discovery issues, and people with more money than they know what to do with beat a path to the inventor's door.

As you might surmise, this image is far from the reality most independent inventors enjoy. It is true that some independent inventors are very successful and have fascinating stories to tell. However, it is worth remembering that the only stories you are going to hear on the news, on the Internet, and through friends are ones about those independent inventors who made it big. The ones who were unsuccessful and blew a lot of money on something that left them with little or nothing to show for it aren't newsworthy because they are not unusual.[2] They also tend not to brag to their friends about their lack of success.

We need to have a frank discussion about what it means to be an independent inventor in the IP Era of the Information Age. What follows is meant to discourage those who have no business pursuing patents and who are only going to end up complaining about the process when the dust clears. The truth is that the odds are stacked against independent inventors, so if you want to be one, you might as well know that going into the game.

One of the biggest hurdles independent inventors face is their own overblown expectations of what a patent on an invention is for and what it can do for them. Many independent inventors see a patent as an end in itself. They tend to view patents from the classical patent mechanics viewpoint and see themselves getting a single spherical patent that acts as a kind of talisman that casts a force field over their invention, killing copycats while spitting out money.

When their patents issue, too many independent inventors do the football equivalent of running to midfield, spiking the ball, and yelling, "Touchdown!" Little do they realize that the people running toward them at this point are not rushing to offer their congratulations. The patent isn't the end game. Without a clear understanding of how the patent is to be leveraged, the independent inventor stands a good chance of watching helplessly as his or her patent gets sucked into a worthless part of IP space and becomes an expensive piece of IP space junk.

12.2 THE CONFLICT OF INTEREST PROBLEM

When a person contacts a law firm, one of the first things the firm wants to determine is how much work that person is going to bring them and whether he or she will be able to pay for it. A patent law firm will also want to know about the technology in question and what organization or company the person represents. The answers to these questions are important because a law firm cannot represent two different parties that have the same or very similar interests—doing so would present a conflict of interest.

Truth be told, when an independent inventor calls a patent lawyer, alarm bells go off in the lawyer's head. The lawyer's first concern arises out of the assumption that the independent inventor has an unsophisticated view of inventions and patenting and thus will likely be a high-maintenance client who needs everything explained to him or her. A second concern is that the inventor does not appreciate the legal costs involved with protecting the invention, not to mention the billable hours the attorney will spend answering all the inventor's questions.

The lawyer's third concern is that the independent inventor, once he or she is taken on as a client, may create a conflict of interest that could preclude the lawyer from taking on a larger, more sophisticated client down the road. Many law firms are not going to risk creating a possible conflict of interest by taking on a small (read: poor) client with a small amount of work only to find themselves with the chance to take on a larger (read: rich) client that could represent a large amount of work in the same IP space.

12.3 THE LACK OF MONEY PROBLEM

No one is ever going to confuse a patent law firm with the Peace Corps or Habitat for Humanity. Patent law firms generally do not go out of their way to provide legal services to clients who don't appear to have the resources to pay. The lack of resources to fund the development of their invention causes most independent inventors to seek relationships with companies that can provide large-scale invention development assistance. Unfortunately, these companies typically have the upper hand in the negotiation unless the independent inventor is fortunate enough to have inventions the company desperately needs.

An alternative for the independent inventor is to try to obtain venture capital to help pay for the development and protection of the inventions and to create a commercial product based on them. However, to do this, the independent inventor is going to need to reinvent himself or herself as a start-up company complete with a solid business plan and all the other things that venture capitalists are going to scrutinize. Often, the independent inventor's best skill—not to mention main interest—is inventing, not running a business. If he or she recognizes this, it may be possible to partner with the right business and venture capitalists to establish a viable start-up. It is a rare individual who has the smarts to do all the jobs necessary to take an idea and make serious money from it without help.

12.4 THE ENFORCEMENT PROBLEM

Most independent inventors are going to find that the cost of enforcing whatever patent rights they can obtain is prohibitive. The cost of initiating and sustaining a patent infringement lawsuit for even a relatively straightforward allegation of infringement costs hundreds of thousands of dollars, and more complicated cases cost many millions. Some lawyers agree to represent independent inventors on a contingency basis, in which they take a percentage (typically 30% to 40%) of the settlement if they win but nothing if they lose. Whether taken on fees or contingency, a lawsuit is a resource-intensive proposition for an independent inventor, not to mention for most companies.

Independent inventors also tend to work under the assumption that a single patent will suffice to provide them with all the legal protection they need to go forth and make deals with companies or start up their own businesses. The reality is that a sophisticated company can invest a great deal of money and effort in improving the technology of that single patent.

For example, if an independent inventor patents a new type of laser, engineers at a competing company might read the patent to learn the technology and then research ways to practice a form of the invention that allows them to avoid the claim limitations of the patent. This ***designing around*** the claims is a completely legitimate endeavor.

A company may also invest time and effort in patenting inventions ancillary to the core invention, such as a method of making the laser mirrors needed to practice the core laser invention and/or methods and apparatuses that employ the core laser invention. This is called "patenting around" an invention. If others do obtain these ancillary patents, then an end user of the core patent will likely need to license the ancillary patents if she or he wishes to use the core invention effectively.

Accordingly, independent inventors need to appreciate that in the IP Era of the Information Age, technology businesses are creating patent portfolios and relying on the collective effect of many pretty good patents rather than on one high-value killer patent. If an independent inventor tries to rely on a portfolio consisting of, say, just one or just a few patents, the inventions and patents need to be truly special.

12.5 CHESTER CARLSON

Chester Carlson invented xerography, which is also called electrophotography. Xerography is an electrostatically based printing process that is still used in photocopiers. Carlson invented xerography as an independent inventor, and his first patent was US Patent No. 2,206,691, which issued on October 6, 1942. Carlson's original xerographic invention was inefficient and not commercially viable; yet, the invention was patented because it met the standards for patentability. Carlson knew to patent his invention even though it wasn't perfect because he also happened to be a patent attorney and knew how to play the patenting game. He was not about to wait until he had perfected his invention before seeking patent protection.

Despite being able to handle the legal side of protecting his inventions himself, Carlson needed money and assistance to develop his invention. He initially had significant difficulty trying to convince companies to work with him to develop and commercialize his invention, and it took him some eight years to find support. Eventually, he teamed up with inventors at two companies, Battelle and Haloid (which became Xerox). They were able to take Carlson's original invention and improve it to the point where it could be commercialized. The result was a portfolio of about forty patents related to xerography—a sizeable portfolio for the 1960s. This portfolio made it very difficult for others to muscle in on the xerography technology.

Observe here that it took substantial effort to take the nascent xerography technology and commercialize it. All of that effort generated commercialization and implementation inventions that were patented to create the formidable xerography portfolio. Carlson and those he teamed up with clearly knew that their patenting efforts needed to be sustained over the long haul as they drove the core invention to commercial viability. Rather than just file a few patents and watch them diminish in value as others patented around them, they stayed ahead of the competition and kept their portfolio up to date and relevant. This is a good, practical example of how a technology business was able to maintain its leadership position in a technology that it invented rather than allowing others to develop and own all the commercialization IP. It also illustrates how an independent inventor may need to combine forces with others to achieve success.

12.6 DR. BRIAN CALDWELL, INDEPENDENT INVENTOR

Brian Caldwell is the sole owner and employee of Caldwell Photographic, Inc., based in Petersburg, Virginia. He invents in the field of optics and, in particular, invents specialized lenses and related equipment for photography and cinematography.

Brian falls into the category of the successful independent inventor. He has patented and licensed a number of his inventions. For example, his US Patent No. 8,289,633, entitled "UV-VIS-IR imaging optical systems," is licensed to Coastal Optics, which sells a UV-VIS-IR 60 mm objective lens based on the patent. The lenses can be used in applications ranging from law enforcement to forensics to fine art.

Brian has also filed a patent application on a focal-reducing attachment lens, which is being made and sold by a company called GBI/Metabones. The attachment is used to modify a camera so that it can operate at a reduced focal length (i.e., a higher F-number) using the existing camera lens. He has also licensed others of his patents.

While Brian doesn't work out of his garage, he does work out of his house. Brian belongs to a breed of independent inventor that works through a technology business he owns and operates essentially by himself. He is not always the sole inventor on his inventions, but he is the driving force behind the main ideas and knows when to call in others to help him squeeze the most value out of the invention.

Brian is successful at inventing and patenting for four basic reasons:

1. He has a high level of technical expertise. He has a PhD in optics from the University of Rochester's Institute of Optics and over 25 years of lens-designing experience. This makes him an expert in his field.
2. He understands that inventions and patents are business tools, so his patenting efforts are very much business driven. This is to say, while he is constantly coming up with new ideas, he knows to spend money on patenting only those ideas that he thinks have a good chance of making money.
3. He knows the market for his inventions because he is actively engaged in the photography and cinematography industries and stays up to date on all the latest products and what people are looking for by way of improvements. He also takes the time to analyze the shortcomings of existing products.
4. He has perhaps the best understanding of his IP space of any inventor on the planet. How did he gain this understanding? At one point in his career, Brian put together and sold a database called LensView™ that included almost every lens-design patent ever issued by the US and Japanese patent offices. This endeavor gave him a nearly one-to-one map of the IP space in which he works. It serves as an IP GPS that lets him see the lens-design landscape and where the "green spaces" (i.e., available IP spaces) exist.

12.7 INDEPENDENT INVENTOR RIOT ACT

Most patent attorneys field calls from independent inventors who have an invention and are desperate to patent it. The independent inventors' expectations and enthusiasm for their inventions are usually extremely high. It can be a difficult task to bring them back down to earth and patiently explain the realities of the patenting process. Most attorneys have a pat monologue that they use to explain the basics of the patenting process to independent inventors who have yet to go through it.

I call my monologue the "Independent-Inventor Riot Act" (IIRA). Like the original Riot Act that was read to disperse crowds of rioters by warning them of the consequences of continuing their riotous behavior, the IIRA warns independent inventors of the consequences of going forward with patenting their inventions without first gaining a realistic understanding of what they are getting into. After all, the people at the IP bazaar who sponsor the advertisements on television that try to suck in inventors with promises of making them rich aren't going to do it.

If I can provide a realistic, if somewhat negative, view of the patenting process and the person is still interested after I'm done, then maybe he or she actually has what it takes to go forward with patenting efforts as an independent inventor. If I dissuade a person from blowing a lot of money based on false hope and woefully unrealistic expectations, then I have helped him or her. An honest attorney does not try to take advantage of a client's naiveté and will live up to the ethical obligation of making sure that any would-be clients understand what they are getting themselves into.

The Independent Inventor Riot Act

1. **Patenting is expensive.** You should be mentally prepared to spend $25,000 for the entire process if you engage a law firm. It might cost less; it might cost more. But $25,000 is the reality check on the potential costs associated with having a patent attorney help you over the long haul that is required to obtain a patent and then maintain it over the course of its lifetime.

2. **Patenting costs are ongoing and annoying.** Patenting is not the kind of activity that involves a one-time cost. There are ongoing costs that become increasingly annoying with time. The ongoing costs can include fees associated with replying to communications from the patent office, hiring a searching agency to conduct a search for prior art, correcting patent-office errors, reporting correspondence from the USPTO, filing documents such as information disclosure statements as required by the patent laws, paying issue fees to get the patent issued, and paying maintenance fees to maintain the patent in force.

3. **DIY is an option.** You are welcome to try patenting your invention via the do-it-yourself route without the assistance of a patent attorney or a patent agent. You can obtain all the forms you need on the Internet. You can even hire an invention promoter like the kinds who advertise on television (and for whom the USPTO has its own standard complaint form). You can file the application in the USPTO by yourself, and in fact you can even do it online and pay the fees with a credit card. There is nothing stopping you from doing everything yourself and not paying a patent attorney a relatively large sum of money to be of assistance. In the short run, it will definitely be less expensive.

---

4. **There are reasons businesses hire patent attorneys.** If you choose to pursue the do-it-yourself route, you should ask yourself this question: Why doesn't everybody with an invention just file his or her own patent applications and get rich? How come technology businesses hire patent attorneys? Remember, there are reasons we hire electricians to wire our houses, plumbers to reroute our pipes, and financial planners to deal with the vagaries of the stock market and investment instruments. Sometimes it makes more sense to hire and pay a professional to do the job right rather than try to do it yourself and have to call the professional later to undo the mess.

5. **Undoing DIY damage.** If you take the do-it-yourself route and then run into problems, don't be surprised if the patent attorney you call does not want to take over your now problematic case. It might not be salvageable, and dealing with substandard legal work poses a potential malpractice risk for the attorney. Here is a great way to create a problem right off the bat using the DIY approach: File your own provisional patent application without any legal advice and without any training in drafting patent applications. Then do nothing for 11 months. Then try to get a patent attorney to prepare the nonprovisional patent application based on the DIY provisional patent application you drafted with the 1-year deadline for conversion only 1 month away.

6. **How will you avoid being a member of the independent-inventor majority?** Only a small percentage of issued patents obtained by independent inventors provide the patent owner with a substantial return. What is on your list of reasons why you and your patent will not fall into this category?

7. **Business tool or hobby?** Decide whether you are pursuing a patent as a hobby or as a business tool. If you are pursuing patenting as a hobby, you need to know that it is going to be an expensive hobby, more like yachting than horseshoes or shuffleboard. If you can't afford yachting as a hobby, then maybe you should reconsider pursuing patenting as a hobby.

8. **If it's a business tool, then you are running a business.** If you are pursuing a patent as a business tool, then you are effectively running a business based on your patent even if you are calling yourself an independent inventor. Maybe you want to try to license your patent to a business. If so, then you are in the licensing business. Maybe you want to have the product made. If so, then you are in the manufacturing business—even if you are outsourcing the design and manufacturing. A business needs a business plan that has a central organizing principle for how the patent is to be leveraged. You should have such a business plan prior to seeking to patent your invention.

9. **Private patent lessons are expensive.** If you plan on hiring a patent attorney with the expectation that he or she will teach you all about patents and the patenting process, you are in essence hiring a private tutor that charges hundreds of dollars per hour to give you private patent lessons. That is an expensive rate for private

lessons for anything. It is a bit like asking Yo Yo Ma to give you private cello lessons. You will learn a lot, but it is going to cost you.

10. **Searching the prior art.** Take the time and incur the expense of having someone conduct a prior-art search that will give you an idea of whether or not the invention could be considered novel and nonobvious. If you are reluctant to pay $500 to $1,500 for a meaningful search of the prior art, then you are likely going to be stunned by the costs associated with actually patenting your invention and maintaining it over its 20-year lifetime.

11. **Prior-art searches are no guarantee.** Just because a prior-art search fails to find any relevant prior art does not mean that no relevant prior art exists anywhere on the planet; it just means that none was found based on the way the search was conducted. It may be that highly problematic prior art related to your invention will show up later and limit the kind of patent protection you can get. This is an inherent risk in the patenting process (recall the prior art uncertainty principle of Chapter 8).

12. **There is definitely prior art for your invention.** Consider that your own bias toward your invention might compromise your objectivity and prevent you from finding the most relevant art. I have seen way too many innovation disclosures that bear the following ridiculous statement under the section entitled "Prior Art": *none found.* I can promise you that there is prior art out there that relates to your invention. The only question is how close it is. By all means do your own search, but also consider having a search conducted by someone who can be objective and brutally honest about the results.

13. **No commercial product does not equal no prior patent.** Just because your invention does not exist as a commercial product does not mean that a clever person in Japan, France, China, or anywhere else people like inventing things has not already thought of it and disclosed it to the public or patented it. Not all inventions become commercial products. Also, patents in a given technology tend to precede the arrival of products in the technology.

14. **Don't ask your patent attorney for business advice.** It is not a patent attorney's job to know about the market for your invention and how to assign a value to a patent that might issue. He or she will usually assume that you did or will do all of that work. Your patent attorney's job is to help you get a patent and serve as your native guide through the USPTO bureaucracy. If you need business advice, seek it from a qualified person.

15. **What is the central organizing principle behind your patenting effort?** You need to be able to articulate a central organizing principle behind why you are seeking a patent (see Chapter 13), and it needs to make sense in view of the costs you are about to incur.

16. **You will need to invest time to create proper documentation.** Be prepared to have to spend significant time and effort documenting the details of your invention and finalizing its form. This will allow another person to review and assess the invention. If you think patenting an invention is expensive, wait until you get the bill for having an attorney make multiple revisions to what was

supposed to be the final draft of the patent application, just so you could include all of your last-minute improvements. If you end up "inventing on the fly" during the patent application process, you may need to take out a second mortgage just to pay for the application. Unlike a conventional business that can operate with invention disclosures that can be fairly terse, an independent inventor needs to flesh out the details of the invention and generate documentation that is heavy on details.

17. **Be prepared to be cross examined by your attorney.** If you hire a patent attorney, you will be answering lots of questions, such as whether you disclosed the invention to anyone; whether you have used it (or a variation of it) and, if so, when you started using it; whether you have been making it and selling it; what your plans for commercializing are; what the related time frame is; and so on.

18. **Learn about NDAs.** Take some time to learn about the importance of nondisclosure agreements and how they are used in connection with inventions and patents.

19. **Seek those who have gone before.** Find and talk to independent inventors who have been through the patenting process and get their take on it. Find at least one who was not successful and one who was so that you can see both sides of the story. Become involved in one or more of the numerous independent-inventor organizations that can easily be found online. They can provide you with a wealth of information and put you in touch with like-minded people who can give you the lowdown on all aspects of the invention and patenting process for a whole lot less money than taking private lessons from a patent attorney would cost you. There are a number of different chat groups that discuss inventions and inventing.

20. **The patenting process takes years.** The pursuit of patents takes patience. It is a marathon, not a sprint. It can take anywhere from 1 to 4 years just to get the first office action. Obtaining the patent may require responding to multiple office actions. You need to be prepared to sustain your enthusiasm and attention for the invention and the patenting process over an extended period of time and in the face of some adversity from the patent office.

NOTES

1. This chapter is adapted from Gortych, J. E. 2003. *Legal lens anthology: Essays on intellectual property*, 66–67. Washington, DC: Optical Society of America. Copyright Optical Society of America and used with permission from the Optical Society of America.

2. John Steinbeck's classic book *East of Eden* (published in 1952) has a character named Samuel Hamilton who "developed a very bad patent habit, a disease many men suffer from." Any would-be independent inventor should read *East of Eden*.

CHAPTER 13

CENTRAL ORGANIZING PRINCIPLES AND PATENT STRATEGIES

13.1 CALLING THE COPS

One buzz-phrase you constantly hear in the IP world is "IP strategy." Much of the talk and writing about IP strategy can be obtuse and hard to follow. What do people mean when they talk about "IP strategy" or a "patent strategy?" One helpful way to think about these concepts is as a plan of action or steps to be taken that support one or more "central organizing principles" (COPs) that explain why IP and patents are being pursued. Defining a COP for a given set of patenting circumstances brings the matter into focus. The patenting strategy is then defined as the steps or actions that need to be taken to support the COP. Writing a COP is like writing haiku. Keep it short and to the point so that it has maximum focus.

Why try to come up with COPs for why you are seeking patents? Because it is not unusual for technology businesses to be pursuing patents in a manner that is not aligned with how they want their patents to serve them from a business viewpoint. If you cannot articulate the fundamental reasons why you are pursuing patents, it is likely that your patenting efforts are not optimally aligned with your business needs.

The best way to see how a COP can bring an IP strategy into focus is to look at a number of basic COPs and their corresponding IP strategies.

13.2 THE NO-PATENTING COP

One approach to patenting that some technology companies adopt is not to bother pursuing patents at all. Given the cost, complexity, time, and resources it takes to implement and operate a patenting system properly, it can be better under certain circumstances just to take the money you would have spent patenting, put it in the bank, and use it to deal with any patent-related problems that might arise in the future.

The basic no-patenting COP can be stated as follows:

> In view of the nature of our business and products, we believe the time, effort, resources, and money we would spend on patents is better spent on investing in other parts of the business, and we accept the risk of and are prepared to address any patent-related issues ex post facto.

What kinds of savings might we be talking about by forgoing patents? Well, assuming that it costs about $25K to prosecute a US patent application, obtain a patent, and maintain that patent over its term of 20 years, we're not talking about an insignificant sum. For a company that patents five inventions a given year, that's $125K annually. If you add up the worker-hours invested in each patent application as well as the worker-hours spent on managing the patenting system in a low- to medium-entropy state, you could easily come up with another $75K per year, to arrive at a "savings" of $200K per year. Over the course of 10 years, that's a savings of $2M. If you add some foreign filings to this mix, you can double or triple this number. This is no small chunk of change for many technology businesses, especially start-ups and small companies.

The no-patenting COP does have a few significant downsides that need to be examined carefully. First, $2M or even $6M is not actually a great deal of money when it comes to patents and patent valuation. $2M is at the low end of the money spectrum when it comes to fighting patent infringement lawsuits. And, if you believe some of the numbers associated with recent purchases of large patent portfolios, the average value of a single patent came out at roughly $700K.

Second, a technology business with no patents is relinquishing the only legal means for obtaining a time-limited monopoly on its products. Such a monopoly can be a way to enhance market share. Third, it leaves the technology business without a patent portfolio that it could leverage against a competitor that threatens legal action based on its own patents.

For a technology business, eschewing patenting altogether can be a bit like deciding not to carry any business insurance. As long as nothing happens, the approach seems simple and wonderful. And if the risk of adverse events is very low, then it can start looking like genius. However, the minute things go wrong, people start wondering what the heck they were thinking by not being insured. The so-called savings of not paying the IP insurance premiums can pale in comparison to the actual costs of dealing with an IP disaster.

So is it reckless for a technology business not to pursue patents? Not necessarily. It really boils down to the question of risk management, the answer to which depends on the products the technology business makes and sells and the dynamics of the IP space. If a technology business wants to take the risk that it will not need patents to be successful and feels as though it can handle any patent-related issues that come its way, then it can be a very sensible business decision. For example, perhaps the products that a particular technology business makes and sells are based on a licensed patent. In this case, someone else has paid all the patenting costs and the technology business just pays a royalty to use the patent. The "IP renter" may be totally justified in feeling that it doesn't want to own any IP. And going patent free while licensing only what you need can be less expensive in the long run.

13.2.1 Trade secret protection as the better option

The no-patenting COP can also work for a technology business that makes products using manufacturing techniques and methods that are not discernible by inspecting or reverse engineering the products. An example of such a technology business might be an optical-component manufacturer. Let's say this manufacturer has a special process whereby the glass is heated in a special manner to anneal out defects and then a special polishing process is used to create super-smooth surfaces. It may be impossible to glean any information about these two special processes simply by inspecting the optical component.

In this case, the special methods, tricks, secret sauces, and know-how are actually better protected as trade secrets. This is especially true in the United States now, as, thanks to the passage of the AIA, one can avoid a charge of patent infringement by showing prior use of an invention if someone ends up later patenting the same invention. The prior use must involve a commercial use or offer for sale and must have occurred more than 1 year before the earlier of the effective filing date or the publication date of the problematic patent.

For the optical widget company and similarly situated technology businesses for which the benefits of the trade secret option outweigh those of the patenting option, the COP can be more specific:

> Because our inventions and innovations can be kept secret by virtue of their undetectability, and because it would be difficult if not impossible to detect infringers of patents on such inventions and innovations because of that same undetectability, we choose to maintain our innovations and inventions as trade secrets rather than to patent them.

13.2.2 Products based on publicly available technology

In another example, it may be that the products that a technology business makes and sells are mostly or entirely based on publicly available technology. Any innovations used to make the products newer, better, faster, smaller, and so on may not be worth patenting based on the history of the business and how the particular industry deals with patents. It may be easier and more efficient just to donate to the public what would be considered minor innovations by simply putting the products out on the market.

Here is the nonpatenting COP for such a technology business:

> Our products and services are not so inventive as to be patentable. We are able to utilize known and unpatented technologies to make products that people want and provide services that people need by having great marketing, attentive customer service, and fair pricing.

An example of such a technology business might be a company that does laser processing of surfaces for industrial applications. If the technology business is buying all the patented equipment from the patent owner, it can't be an infringer of the equipment due to the doctrine of patent exhaustion. (Wouldn't it be interesting if a company could sell you a tool and then sue you for patent infringement when you tried to use it for its intended purpose?) If the laser system and the methods used to perform the laser processing are straightforward, then it can be that there is simply no motivation on the part of the business to try to patent anything.

This approach works well for companies that have achieved solid brand recognition and a solid market share by virtue of making great products and knowing how to keep their customers happy.

This, by the way, leads us to one of the more important fundamental truths about patents:

> **Patents can be terrific business tools when properly wielded, but they can't make up for lousy products and dreadful customer service.**

So, the good news here is that there are COPs that support a no-patenting approach to patents, which means that there is no need for an IP strategy, right? After all, if a company isn't going to patent anything, what is there left to do from an IP viewpoint?

Once again, things are never as easy as they might look. The IP strategy for the no-patenting COP has several key facets. The first involves performing a proper analysis of the nature and content of the IP involved to make sure this COP works for the business and its unique risks. It may be that only a segment of the business is amenable to the no-patenting approach. The second and third facets cannot be overemphasized: (a) properly maintaining and managing the trade secrets, and (b) managing the maintenance and flow of confidential information.

13.2.3 The ill-conceived no-patenting COP

Many technology businesses choose the no-patenting approach for the wrong reasons. Here is another, more honest version of a no-patenting COP that applies to some technology businesses:

> **We are too busy, under-resourced, and lacking in the institutional knowledge and inclination to operate an internal patenting system to protect our inventions.**

This "we're too busy" COP is the unstated reason on which many technology businesses skate for years without bothering to patent anything significant in an IP space that screams for at least some level of patenting. Usually, the odds catch up with such companies. Sooner or later, IP disaster strikes, and these companies find themselves unable to stop copycats and competitors or embroiled in a patent-infringement lawsuit with no patents of their own to use as leverage.

The key point here is that the decision to forgo patenting must be an *informed* one, not an ill-conceived one. Informed people run the numbers, understand their products and the IP involved, evaluate the pros and cons of patent protection for the technology at hand, and make calculated choices. The aforementioned optical component manufacturer and similarly situated widget makers that choose not to patent their manufacturing methods but to maintain them as trade secrets are making an informed choice. The company that operates based on the ill-conceived no-patenting COP is abdicating its responsibility and putting its fate in the hands of others.

13.3 THE BOTTLENECK IP COP

Technology businesses that develop truly core or "bottleneck" inventions are in an enviable position. Bottleneck inventions are by definition so fundamental that essentially everyone in the industry needs to use them to practice the technology. The portfolio of xerography patents created by Chester Carlson and his co-workers is a good example of inventions covered by bottleneck patents. If you wanted to practice xerography back in the day, you needed to go through Chester Carlson and his posse of inventors.

A more recent example of a bottleneck invention is the digital mirror device, or DMD. Texas Instruments pioneered the development of DMDs and has a sizeable portfolio of patents around DMDs and their myriad applications. DMDs work by controlling the positions of very small mirror elements arrayed in a particular shape. A light beam illuminates the mirror elements, which are rapidly adjusted (e.g., several thousands of times per second) by a computer processor. The "on" and "off" states refer to the mirror being angled such that light is reflected toward and away from the viewer, respectively. By controlling the amount of time a given mirror element remains in the "on" state, you can control the amount of light reaching the viewer. This allows the mirror array to be used to create images that can then be displayed. Certain types of televisions and displays utilize DMD technology.

Texas Instruments makes a huge amount of money from licensing its DMD patents. A quick search of Texas Instruments patents using the search term "DMD" yielded over 750 hits. The core DMD IP includes Texas Instrument's US Patent Nos. 4,441,791; 4,566,935; and 4,662,746. Figure 13.1 is the first page of the '746 patent, entitled "Spatial light modulator and method," which shows an electrostatically deflectable, spatial-light-modulator pixel in the form of a flap 30 that rotates about a hinge 34. The device is supported by a substrate 22 and includes a spacer 24.

The COP for a technology business that has such core/bottleneck/pioneering technology and wants to patent it so that it can leverage it to the hilt can be stated as follows:

> **If the core technology we invented is ever adopted, we want to make sure that anyone and everyone who wants to use it has to come through us.**

Texas Instruments realized that the DMD inventions were not just improvement inventions on someone else's stuff. It realized that it had really and truly invented bottleneck IP and leveraged it to its advantage, especially in televisions and projectors based on digital light projection (DLP) technology (DLP is a trademark of Texas Instruments.).

So with this COP, the patenting strategy is straightforward: If you want the industry to come through you, you need to create a patent portfolio that includes

United States Patent [19]

Hornbeck

[11] **Patent Number:** **4,662,746**

[45] **Date of Patent:** **May 5, 1987**

[54] **SPATIAL LIGHT MODULATOR AND METHOD**

[75] Inventor: **Larry J. Hornbeck,** Van Alstyne, Tex.

[73] Assignee: **Texas Instruments Incorporated,** Dallas, Tex.

[21] Appl. No.: **792,947**

[22] Filed: **Oct. 30, 1985**

[51] **Int. Cl.⁴** G02B 26/02; G02B 00/00; H01S 3/00

[52] **U.S. Cl.** 350/269; 332/7.51; 350/320

[58] **Field of Search** 350/632, 360, 487, 486, 350/6.6, 269

[56] **References Cited**

U.S. PATENT DOCUMENTS

3,600,798	8/1971	Lee	350/320
3,886,310	5/1975	Guldberg et al.	350/360
3,896,338	7/1975	Nathanson et al.	358/60
4,229,732	10/1980	Hartstein et al.	358/233
4,566,935	1/1986	Hornbeck	350/360
4,596,992	6/1986	Hornbeck	350/360

Primary Examiner—John K. Corbin
Assistant Examiner—Vincent J. Lemmo
Attorney, Agent, or Firm—Carlton H. Hoel; Leo N. Heiting; Melvin Sharp

[57] **ABSTRACT**

An electrostatically deflectable beam spatial light modulator with the beam composed of two layers of aluminum alloy and the hinge connecting the beam to the remainder of the alloy formed in only one of the two layers; this provides a thick stiff beam and a thin compliant hinge. The alloy is on a spacer made of photoresist which in turn is on a semiconductor substrate. The substrate contains addressing circuitry in a preferred embodiment.

18 Claims, 23 Drawing Figures

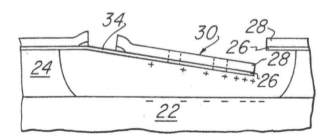

Figure 13.1 First page of Texas Instruments' US Patent No. 4,662,746.

not only the core patents but improvement patents and commercialization implementation patents.

To be sure, lots of companies have patents that include DMDs. But Texas Instruments has the core patents, not to mention a large number of improvement and commercial implementation patents. For example, Figure 13.2 shows the first page of US Patent No. 5,247,180, which was filed in December 1991 and is directed to a "stereolithographic apparatus and method of use."

A stereolithography apparatus is used for rapid prototyping and forms three-dimensional (3D) objects by building up thin layers of a UV-curable resin. Stereolithography is presently a very hot technology and was mentioned by

United States Patent [19]

Mitcham et al.

US005247180A

[11] **Patent Number:** **5,247,180**

[45] **Date of Patent:** **Sep. 21, 1993**

[54] **STEREOLITHOGRAPHIC APPARATUS AND METHOD OF USE**

[75] Inventors: **Larry D. Mitcham, Temple; William E. Nelson, Dallas, both of Tex.**

[73] Assignee: **Texas Instruments Incorporated, Dallas, Tex.**

[21] Appl. No.: **814,859**

[22] Filed: **Dec. 30, 1991**

[51] Int. Cl.⁵ **B29C 35/08**

[52] U.S. Cl. 250/492.1; 264/22; 425/174.4; 346/108; 346/160

[58] Field of Search 250/492.1; 264/22; 425/174, 174.4; 346/108, 160; 364/522; 356/121, 375

[56] **References Cited**

U.S. PATENT DOCUMENTS

5,011,635	4/1991	Murphy et al.	250/492.1
5,031,120	7/1991	Pomerantz et al.	364/468
5,049,901	9/1991	Gelbart	346/108
5,059,359	10/1991	Hull et al.	250/492.1
5,123,734	6/1992	Spence et al.	425/174.4
5,164,128	11/1992	Modrek et al.	250/492.1

OTHER PUBLICATIONS

3D Systems, Inc. brochure.

Primary Examiner—Paul M. Dzierzynski
Assistant Examiner—Kiet T. Nguyen
Attorney, Agent, or Firm—Julie L. Reed; James C. Kesterson; Richard L. Donaldson

[57] **ABSTRACT**

A stereolithographic apparatus is disclosed comprising a vat, a working surface in the vat, an elevating mechanism for controlling the level of liquid within the vat relative to the working surface, an illumination source for emitting radiation, and an area array deformable mirror device. The illumination source emits radiation which is operable to harden a stereolithographic liquid, while the deformable mirror device is operable to reflect the incident radiation onto the surface of the liquid. The deformable mirror device can harden an entire lamina of liquid in one brief exposure interval increasing throughout without sacrificing resolution.

16 Claims, 3 Drawing Sheets

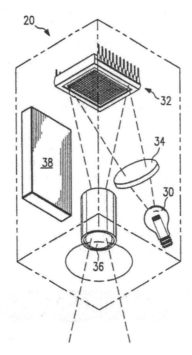

Figure 13.2 First page of Texas Instruments' US Patent No. 5,247,180.

President Obama in his 2013 State of the Union address. Texas Instruments did not invent stereolithography—Charles Hull did in 1986, and he patented his original version of the invention as US Patent No. 4,575,330. But Texas Instruments patented its DMD-based version of stereolithography not that long after Hull's original invention.

So Texas Instruments didn't just invent the basic DMD technology and forget about it. It drove it all the way down to the commercial level, just like Chester Carlson did with xerography. It also patented a number of key inventions directed to important applications that make use of the core DMD technology.

13.3.1 Patenting beyond the core technology

A problem with just patenting the core technology and not trying to own the improvement and the commercial implementation patents as well is that the value of a patent portfolio based only on core or bottleneck IP will diminish over time.

Think about what happens anytime someone comes up with something of value: People rush in and want a piece of the action. This is true for real estate, gold mines, stocks, fashionable clothes, modern art, cars, jewelry, and patents. It is human nature not to let others have all the wealth for themselves. In the corporate context, it is bad business to let other companies have the entire market share without even trying to compete.

Imagine you've created a new IP space by inventing core IP. That whooshing sound you hear soon afterward is everyone rushing in to get a piece of the action. If you just sit there with your initial invention while basking in the glory of having invented it first, you will quickly be surrounded by others with improvement and commercialization inventions based on your core technology (that is, your core patents will be patented around). A fair number won't even acknowledge that you invented the core technology. If the core technology is valuable, then it is virtually guaranteed that someone else will try to rip it off, figuring he or she will just pay the damages if caught. By then he or she will have pocketed a fair amount of money.

It may be that you or others will need to use the improvement and commercial implementation patents that end up surrounding your core patents. These improvement and commercial implementation patents can end up representing layers of tax that are added onto your core IP if you or someone else wants to use the core IP.

So the bottleneck IP patenting strategy involves more than just patenting core inventions. If you want everyone to have to come through you, you need to own the IP space outright.

13.4 THE COMMERCIALIZATION COP

A rather amazing aspect of patenting is that you don't have to be the original inventor of a technology to obtain patents in that technology. So far we have discussed the benefits of holding core patents as well as improvement and commercial implementation (or "commercialization") patents, but not all technology businesses own patents in each of these categories. In fact, it is the rare technology business that does. Many technology businesses focus on taking a core technology invented by someone else but not yet commercialized and driving it to commercialization.

Thomas Edison, for example, is well known for "inventing" the light bulb. But the truth is that light bulbs were around long before Edison started inventing in the light bulb IP space. Edison's proper claim to fame is that he invented a *commercially viable light bulb* based on a carbon filament in a vacuum. He did not invent or own the core IP in the light bulb IP space.

So one approach to patenting for a technology business is to take a core technology invented by someone else that has not yet been commercialized and drive it to commercialization. The COP reads something like this:

> **We did not invent the core technology, but we will pursue improvement and commercialization innovations and patent them to make the technology commercially viable.**

Figure 13.3 is the cover page of US Patent No. 5,307,410 to Bennett. The Bennett patent is assigned to IBM, and the last time I checked IBM still owned it. The Bennett patent is a pioneering patent and one of the core patents in a branch of quantum cryptography technology called *quantum key distribution* or *QKD*. QKD involves sending single-photon light pulses from a sender to a receiver and encoding the light pulses with a randomly selected phase or polarization. The encoded single photons constitute "qubits," which is an abbreviation of the phrase "quantum bit."

If an eavesdropper attempts to intercept and measure the state of the single-photon pulse and then resend it to the receiver, that measurement will cause the error rate of the transmission to increase and thereby reveal the presence of the eavesdropper. The error introduced by the eavesdropper measurement has to do with the quantum nature of the single-photon pulses.[1]

Claim 10 of the Bennett patent is reproduced here.

10. A method for generating cryptographic key information comprising the steps of:
generating and sending a plurality of dim light pulses of coherent light of intensity of less than 1 expected photon per dim pulse spaced apart in time,
setting the phase of each of said plurality of dim light pulses, said phase chosen randomly from a plurality of predetermined values in response to said random numbers, and
recording the respective phases of said plurality of dim light pulses as a function of time.

United States Patent [19]

Bennett

US005307410A

[11] **Patent Number:** **5,307,410**

[45] **Date of Patent:** **Apr. 26, 1994**

[54] **INTERFEROMETRIC QUANTUM CRYPTOGRAPHIC KEY DISTRIBUTION SYSTEM**

[75] Inventor: **Charles H. Bennett,** Croton-On-Hudson, N.Y.

[73] Assignee: **International Business Machines Corporation,** Armonk, N.Y.

[21] Appl. No.: **66,743**

[22] Filed: **May 25, 1993**

[51] **Int. Cl.⁵** **H04L 9/08; H04L 9/22; H04B 10/18; H04B 10/20**

[52] **U.S. Cl.** **380/21; 380/44; 380/59; 359/112; 359/118**

[58] **Field of Search** 380/21, 43, 44, 33, 380/54, 59; 359/112, 118

[56] **References Cited**

U.S. PATENT DOCUMENTS

4,866,699	9/1989	Brackett et al.	370/3
4,879,763	11/1989	Wood	455/607
4,903,339	2/1990	Solomon	455/612
4,965,856	10/1990	Swanic	455/617
5,140,636	8/1992	Albares	380/54
5,191,614	3/1993	LeCong	380/49

Primary Examiner—Gilberto Barrón, Jr.
Attorney, Agent, or Firm—Robert M. Trepp

[57] **ABSTRACT**

An apparatus and method for distributing cryptographic key information is described incorporating a quantum channel for conveying dim and reference light pulses, a timing channel, a source of coherent light pulses, beamsplitters, a random number generator, a phase modulator and a memory for recording the phase of transmitted dim light pulses. A cryptographic key receiver is described incorporating beam splitters, a random number generator, a phase modulator, a detector and a memory for recording the phase of received dim light pulses. The invention overcomes the problem of distributing fresh cryptographic key information between two users who share no secret information initially.

10 Claims, 3 Drawing Sheets

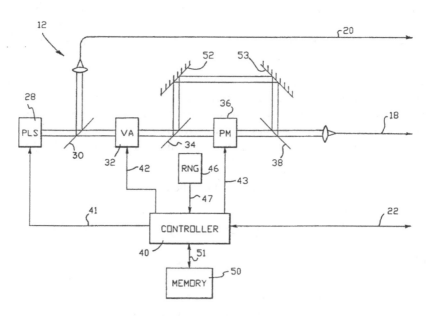

Figure 13.3 Cover page of IBM's US Patent No. 5,307,410.

Note that this claim contains no performance limitations. For a QKD system to be useful, it needs to send signals across appreciable distances over which people might communicate, perhaps between two buildings or even two cities. The claim also contains no limitations on how to package the system to make it tamperproof or how to transmit a message. This is a very good example of a core (or "pioneering") patent claim that we discussed in Chapter 7: It simply captures the basic method that Bennett invented without heaping on unnecessary limitations as to the specific forms the apparatus may take, whether the signals are sent on an optical fiber or through free space, what kind of detector is used, whether you need a laser, and so on.

Another thing to note about this patent is that you can't purchase a commercial QKD system from IBM like you can an IBM mainframe. That is not unusual; IBM patents many inventions that it does not use or embody in its products. In fact, IBM manufactures and sells and leases (licenses) patents as a separate line of business and is very good at it—to the tune of billions of dollars per year in licensing revenue.

So if a commercial QKD system is to be developed, who is going to do it? It is not likely to be IBM, since they have yet to do so despite holding a core patent as well as some ancillary QKD patents. In this case, it will come from one or more companies that want to take the risk of driving the technology to commercialization.

Enter MagiQ Technologies, a small company based in Somerville, Massachusetts. MagiQ adopted a strategy based on driving the core QKD technology to the point where a commercial QKD system could be manufactured and sold. To do so, it faced the daunting task of innovating to address a huge number of practical implementation and commercialization challenges. One challenge was to increase the distance over which the QKD could be used (with about 100 km believed to be a natural limit). Other challenges related to automating the calibration of QKD stations, automating synchronization between stations, automating system boot-up methods, and increasing security with regard to system hardware.

Over the course of 5 years or so, MagiQ filed close to one hundred patent applications on inventions directed to commercialization patents for the QKD system. To date, it has received about fifty patents related to QKD system and quantum cryptography.

The evolution of the quantum-key distribution IP space is quite interesting. The big bang that created the QKD IP space arguably started with the Bennett patent. Soon after, other patents started filling the IP space even though commercialization was not even on the horizon. In fact, even though the market for QKD systems has yet to materialize, patenting continues to expand the IP space. Not all the patenting is commercially oriented. Some patents are directed to improvements upon or simply different forms of QKD systems. Technology businesses are spending money on patenting in the QKD IP space because they perceive that one day there will be value in owning a piece of the space even though the options and demand for purchasing QKD systems remain quite limited.

As it waits for the market to arrive for QKD systems, MagiQ Technologies has built its portfolio in the QKD space to a state of significance, and that significance can be stated in the following COP:

> **If the market for QKD systems takes off, we want a seat at the IP table.**

13.5 NEW PRODUCTS FROM AN OLD TECHNOLOGY COP

Some technology businesses improve an old technology they did not invent to make new products rooted in the old technology. The COP that such a technology business might adopt can be expressed as follows:

> **We did not invent the basic technology underlying the product, but we can develop a new and improved version of the old product and patent innovations directed to improvements to each component, as well as to combinations of the improved components.**

One example of an old technology that has undergone a resurgence in recent years is 3D movies and 3D television, whose technologies have been around for more than 50 years. One of the earliest 3D movies, *The Power of Love,* was released in 1922.[2] Audience members watched the movie through a pair of glasses having one red lens and one cyan lens, which served to merge the offset images of the different colors on the screen and thereby create the 3D effect. 3D movies were popular in the 1950s and again in the 1980s, and the 3D craze seems to be upon us once again thanks to improvements in computers and device technology.

The basic approach to creating a 3D display via either a movie screen or a television has remained generally unchanged since its development. A basic 3D display system requires:

1. A viewing screen upon which stereoscopic images can be displayed
2. A polarization switch that controls the polarization of the light from the screen
3. Eyewear to allow a viewer to see the stereoscopic images
4. A controller that switches the polarization of the images or the transmission of the lenses of the eyewear

An example of the core IP for a 3D television is US Patent No. 2,508,920 from 1950, entitled "Television system" and invented by R. D. Kell. The cover of the Kell patent is shown in Figure 13.4. It is worth taking a closer look at what is going on in this patent to see this particular COP in action.

Figure 1 of the Kell patent shows a television transmitter that receives light from an object that travels along two left (L) and right (R) optical paths through respective vertical and horizontal polarizers 32 and 34. The vertically (V) and horizontally (H) polarized light is combined using a half-silvered mirror 24, which directs the light to a rotating drum 38, shown in close-up in Figure 2. The rotating drum 38 is rotated by a motor 40 and has a stationary reflecting surface 36 located inside. The rotating drum has H and V sections that transmit H and V polarized light. This causes the object light from the L and R optical paths to be transmitted alternately to the television camera tube 10 that has a light-responsive electrode surface 11. The result is that the television transmitter

May 23, 1950   R. D. KELL   2,508,920

TELEVISION SYSTEM

Filed Dec. 12, 1945

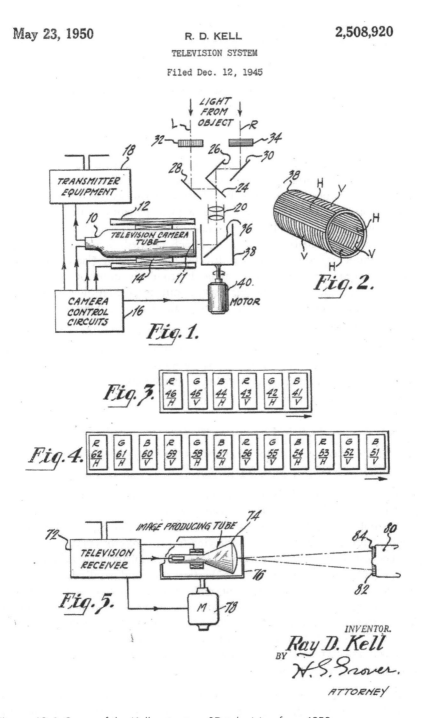

Figure 13.4 Cover of the Kell patent on 3D television from 1950.

transmits stereoscopic images to a television receiver 72 (which is shown in Figure 5 of the patent). The television receiver 72 displays the offset images transmitted by the television transmitter. A rotating drum 76 in front of an image-producing tube 74 polarization modulates the offset images. Glasses 80 are worn by the observer, where lenses 82 and 84 have V and H polarizations, respectively. For a field repetition rate of 60 Hz, each eye sees thirty V images per second and thirty H images per second. This gives rise to the 3D stereoscopic effect.

As 3D technology advanced, people made improvements to the mechanical aspects of the earlier 3D television technology. Kell's mechanical polarization switch has been replaced by liquid crystal switches (see, for example, US Publication No. 2013/0135285, which has a good "patentese" title: "Polarization switching device, driver of polarization switching device, and method of driving the same"). Similarly, US Patent 3,373,568, entitled "Stereoscopic apparatus having liquid crystal filter view," is directed to 3D-viewing glasses wherein the lenses are liquid crystal filters that can be electronically turned on and off rapidly by a view controller. This allows for fast switching between the left and right eyes when viewing a television on which left-eye and right-eye stereoscopic pictures can be displayed.

The patent strategy for this COP involves identifying the key aspects of the older technology and then improving and patenting the various key aspects. In the case of 3D technology, the improvements lie in combining newer components to achieve improved 3D systems.

13.5.1 Example company: RealD, Inc.

A technology business that appears to be employing this COP for 3D technology is RealD, Inc., located in Beverly Hills, California.

RealD has pursued and obtained patents for new and improved components of the basic 3D display technology to create new 3D display products. For example, its US Publication No. 2008/0143065, entitled "Combining P and S rays for bright stereoscopic projection," is directed to the projector component of a 3D display system that provides enhanced brightness of the stereoscopic images. Its US Patent No. 7,679,641, entitled "Vertical surround parallax correction," is directed to the controller component of the 3D display system and takes advantage of progress in computers and software to improve the quality of 3D movies. Moreover, its US Patent No. 7,528,906, entitled "Achromatic polarization switches," is directed to the polarization switch component of the 3D display system and seeks to improve the color performance from the ancient 3D Era that employed mechanical switches.

RealD's US Patent No. 7,583,437, entitled "Projection screen with virtual compound curvature," is directed to the screen/display part of the 3D system.[3] The new-fangled projection screen is made up of a plurality of angled planar reflective surfaces. In addition, RealD's US Publication No. 2008/0151370, entitled "Method of recycling eyewear," seeks to patent, as the titled suggests, a method of recycling used 3D eyewear.

13.5.2 Other examples of the new products/old technology COP

A close look at the spectrum of technology businesses reveals that many of them utilize this COP. Automobile companies are a prime example. The automobile was invented in the early 1900s and the basic parts are the same. While new parts have been invented, the old parts (tires, windshield, motor, transmission, lights, etc.) have been steadily updated and improved with time, as we saw back in Chapter 8 when we discussed automobile technology to illustrate the fractal nature of innovation.

Another good example of this COP involves the sewing machine. The name "Singer" is usually associated with the modern sewing machine. But Isaac Merrit Singer did not invent the sewing machine. Rather, in the mid-1800s, he made some key improvements to existing sewing machines, including a foot treadle that powered the needle to move up and down. Equally if not more importantly, he teamed up with a fellow named Edwin Clark, who used innovative marketing techniques (including allowing for payment for a machine in installments) that served to popularize the sewing machine with women and make it a household item.

However, even Singer ran into patent problems. He was sued by Elias Howe for patent infringement for the lockstitch mechanism Singer used in his sewing machine. While Singer lost the lawsuit, it did not prevent him from growing his business into a multinational corporation and making a huge fortune for himself. Interestingly, even though Howe had a key patent on one aspect of sewing machine technology, his efforts to grow his sewing machine business never got the kind of traction Singer did by being smart enough to team up with what turned out to be one of the biggest marketing gurus of the Industrial Age.

13.6 ASSERTIVE LICENSING COP

There are some technology businesses whose sole business plan is to derive most or all of their income from patent licensing. The COP for assertive licensing is as follows:

> **We want to obtain patents and compel others to license them.**

Here, "obtain patents" can mean developing the technology and patenting it or obtaining the patents from the IP bazaar (e.g., through a patent aggregator, other companies, online auctions, etc.). Companies that obtain patents with the sole intention of compelling licenses are referred to as nonpracticing entities (NPEs) or more pejoratively as "patent trolls." I prefer the less judgmental term "assertive licensor," though the double entendre of the term "offensive licensor" might in some cases be more apt.

The IP strategy that supports this COP depends at its basic level on obtaining patents that actually matter to someone else. That is to say, the COP depends on obtaining patents on inventions that others actually need and want to use and that cannot be readily designed around. The only way to compel a company (sometimes called the "target licensee") to license a patent is to make sure that it otherwise has no choice but to infringe it. Actually, for some assertive licensors, there has only to be a whiff of potential infringement for them to approach a target licensee. Some target licensees will do what it takes to avoid litigation, even if they believe they are not infringing, and thus can be compelled to take a license.

Recall that to capture a large number of infringers, a patent's claims have to be suitably broad, which also makes them more readily invalidated. Some assertive licensors walk the fine line between obtaining patents that others will infringe while ensuring that the claims can withstand the inevitable invalidity challenge. For others, walking that fine line isn't their style, and they have no problem enforcing overbroad claims with a heavy hand.

As you might imagine, a key aspect of this IP strategy is that the technology business is prepared to litigate to enforce its IP rights. The term "compel" in the assertive licensor's COP says it all. This COP is not for the faint of heart. It is more like one that Darth Vader would use and enjoy. Where some technology businesses are litigation averse, assertive licensors are litigation inclined if not predisposed. The more sophisticated assertive licensors perform a great deal of prelitigation strategizing to make sure any patent infringement lawsuit that is filed proceeds along a preferred path. This work includes anticipating all the invalidity arguments the target licensee could use in an attempt to invalidate the patent and shoring up any weaknesses if possible. It also requires a huge investment of time and money. Patent litigation is intense, expensive, and confrontational, and usually involves high stakes.

This IP strategy also requires investigating companies and their business activities to discern whether infringement is possibly taking place. This investigation includes identifying target licensees, picking which ones to go after, and determining what kind of licensing deal to offer.

A good example of a technology business that operates as an assertive licensor is Rambus, Inc., headquartered in Sunnyvale, California. Rambus designs, develops, and patents technology relating to chip-interface technology. In 2009, it acquired Global Lighting Technologies, Inc., thereby expanding its interests beyond semiconductor technology and into advanced lighting and optoelectronics. The acquisition included eighty-four patents mainly directed to LCD backlighting technology used with displays for computers, smartphones, and so on.

The story of how Rambus ended up becoming an assertive licensor depends on the storyteller. One story has it that the assertive licensor COP was forced upon Rambus by uncooperative and hostile memory-manufacturing companies during discussions relating to technology standardization. Another story goes that Rambus was generally hyperdefensive and unwilling to work in accordance with the standards organization because it wanted to leverage its patents optimally.

Either way, Rambus decided that it needed to develop and license its chip-interface technology actively. Given that there was (and presumably still is) a great deal of animosity between Rambus and others in the memory business, it is not surprising that many of its licensing efforts ended up in litigation.

According to the USPTO database, Rambus has close to 1,000 US patents. The history of litigation involving Rambus is legion and spans fights with the Federal Trade Commission and the International Trade Commission, accusations of antitrust violations, and patent infringement lawsuits against memory-chip makers. Both the federal courts and the USPTO have invalidated many of Rambus's key patent claims. What is less publicized is the fact that Rambus receives about 85% of its total revenue from licensing and until lately was making tens of millions of dollars a year in royalties. A partial list of patent licensees includes the likes of Advanced Micro Devices, Inc., Fujitsu Limited, Intel Corporation, Panasonic Corporation, NEC Corporation, Oki Electric Industry Co., Ltd., Renesas Technology Corporation, and Toshiba Corporation.

As of this writing, Rambus has taken a substantial financial hit, having lost a huge antitrust lawsuit it brought against chip makers Micron Technology Inc. and Hynix Semiconductor Inc. Rambus alleged that the two companies had conspired to fix the prices of memory chips to the detriment of the chip market and of Rambus in particular. As Rambus was seeking billions of dollars in damages, the loss was seen by the markets as a major blow to the company. That said, a more recent district court case against SK Hynix was decided in Rambus's favor; the court ruled that the Rambus patents at issue were valid and infringed and that Rambus was entitled to a reasonable royalty.

13.6.1 Academically based assertive licensors

A less conspicuous type of assertive licensor is the higher educational academic institution with an active technology-transfer office. The Bayh–Dole Act[4] of 1980 addressed the subject of patent rights to inventions developed with government funds. Since most higher educational research-based academic institutions run on government grants, the inventions they generate fall under the Bayh–Dole Act. The act essentially gives inventors the rights to their inventions, with the government reserving rights in the form of a royalty-free license to use the invention if it so chooses.

Most academic institutions, however, require that their employees (and in particular their professors) assign their invention rights to the institution. The institutions can end up making quite a bit of money by patenting and licensing the inventions while the government gets a royalty-free right to use patents. These efforts are often controlled and carried out by the institution's technology-transfer office, which is charged with commercializing the inventions. This commercialization has at its core the licensing of any patents the institution owns. The Bayh–Dole Act requires that the inventors share in any licensing royalties the institution receives.

The engagement of academic institutions in assertive licensing is especially interesting because colleges and universities have historically been seen as promoting academic freedom and the free exchange of information—but not so much anymore. In some cases, academic institutions can get quite pushy in their offensive licensing efforts. There have been a few high-profile patent cases involving academic institutions.

CASE STUDY: The Rochester Patent

One academically based patent case involved the University of Rochester and its US Patent No. 6,048,850, entitled "Method of inhibiting prostaglandin synthesis in a human host" and directed to cyclooxygenase (COX) inhibitors. Let's call this patent the "Rochester patent."

The Rochester patent issued on April 11, 2000. That very day, the University of Rochester sued G. D. Searle & Company and related companies including Pfizer, which makes Celebrex, a COX-2 inhibitor. Rochester alleged that Pfizer and other drug companies were infringing the Rochester patent. COX-2 inhibitors are a form of a nonsteroidal anti-inflammatory drug (NSAID). But maybe the more important thing to know is that sales of Celebrex were reported at $2B in 2006 and $2.55B in 2011. The total market for such anti-inflammatory drugs was estimated at $5B in 2003. At stake for the University of Rochester was billions of dollars in royalty payments for anti-inflammatory drugs based on the COX-2 inhibitors. Rochester spent some $10M pursuing the case.

Naturally, the defendants asserted that the patent was invalid. At first blush, supporting this assertion would appear to be a daunting task. The Rochester patent is some sixty pages long and is replete with impressive figures and pharmaceutical jargon. Viewed as a spherical patent, it looks very impressive. However, when the district court measured the Rochester patent, it found that it did not satisfy the written description requirement. The patent wavefunction for the Rochester patent thus collapsed to an invalid state, which in practical terms means that the University of Rochester lost. Rochester appealed the lower court decision to the CAFC, which upheld the decision of the lower court.[5]

The problem the two courts had with the claims of the Rochester patent was that they were so-called "reach-through" claims.[6] A reach-through claim tries to protect something that the invention could be used to make in the future even though that something has not yet been invented. Since the Rochester patent did not describe how the invention could be used to make specific drugs that would presumably be subject to royalty payments, the patent was deemed to have fallen short of the written description requirement. This means the court found that the inventors were not in possession of the claimed invention because they did not describe the claimed invention in the patent.

The commercialization efforts of technology transfer offices at academic institutions can end up resembling Rambus's assertive licensing approach more than they might like to admit. After all, if a university ends up spending substantial amounts of money developing various technologies and then patenting them, there is the clear expectation that the effort has at least to pay for itself—if not make a profit. If a technology business is infringing one of those patents, then the university might have no choice but to go after the infringer if it refuses to license. Like Rambus and other assertive licensors, academic institutions don't actually manufacture anything, so they are not interested in cross licensing. This means that if a target licensee refuses to license, the university may be forced to litigate. This presents a bit of a dilemma for those academic institutions that would prefer not to look like a patent troll.

On the other hand, academic institutions also use their patents to spin out companies in their efforts to commercialize technology. They license the patent to the company, which is usually run by one or more of the university-affiliated inventors. In this kind of situation, if the spin-out company somehow got itself into trouble by infringing another patent, cross licensing might not be out of the question. Here, the assertive licensor COP is modified as follows:

> **We want to patent technology invented by our researchers and commercialize it by licensing it to a spin-out company run by the researchers/inventors.**

The key part of the IP strategy behind this spin-out COP is vetting the inventions by performing serious market analysis to make sure that the time and effort that go into the research and development and then the subsequent patenting of the technology are truly worthwhile.

As axiomatic as the preceding sounds, many academic institutions are horrible at this part of the strategy and instead patent first, leaving the licensing angle to be figured out later. This "ready–fire–aim" approach to licensing is very hard to get to work and is very expensive. This approach is particularly puzzling for academic institutions that have a business school that could help perform a business analysis of the invention. Effective academic patenting requires that there be someone with clout in the academic patenting system that can impose a reality check on the patenting process. This is particularly important when dealing with professors, who know a great deal about their subject matter but often know much less about business, patents, and commercializing technology.

One problem with trying to commercialize academically generated technology is that it sometimes is so cutting edge that it is eons away from commercialization. Core technology that opens up a new IP space is a wonderful thing, and it would be a shame not to protect it. But as we discussed, commercializing core technology that is far from commercialization typically involves a huge development effort (and the attendant patenting effort and expense) followed by the detailed commercialization work (and attendant commercialization patenting effort and more expense) to get to a product that lots of people will want to buy.

As our discussion of the core patent COP emphasized, the core academic invention will likely need to be followed up by enough patents to create a portfolio that protects the core and creates ownership of the IP space. This will likely be achieved by the spin-out company, which will need the necessary funds to pursue the patents in parallel with the development and commercialization work. Note that the one-patent-for-one-product approach doesn't work with nascent core technology because it is virtually impossible to anticipate all the parts and processes that will be necessary to make a future commercial product.

13.6.2 Assertive licensing by proxy

It is not just technology businesses and academia that are involved in assertive licensing. Many IP consulting companies are heavily involved in the licensing game.[7] A good example is General Patent Corporation, an IP management firm (not a law firm) based in Suffern, New York. Its business model includes assisting its clients with IP licensing and enforcement on a contingency basis, which means if the clients don't win, General Patent Corporation doesn't get paid. General Patent Corporation is generally known for taking on the smaller client and assisting in assertive licensing to larger companies, including such giants as IBM and Motorola. If litigation is required, it engages a law firm and manages the process on behalf of its clients.

By working on a contingency basis, General Patent Corporation allows smaller businesses and individuals to obtain the leverage they generally don't have. To be successful, however, General Patent Corporation takes on only a few select cases out of hundreds of submissions; the assertive licensing proposition has to have good prospects or the company runs the very real risk of not ever getting paid.

Another IP consulting company that helps patent owners assertively license their patents is IPValue, which in the early 2000s was busy trying to assertively license British Telecom's patent relating to hyperlinking. The emergence of these so-called licensing boutiques reflects an increasing emphasis on monetizing patents. It is not only big companies that are seeking to monetize but also small players with patents who wish to try their hand against the larger companies. Just as some law firms will take a case for a small client against a large client on a contingency basis, IP consulting companies like General Patent Corporation will take the risk when the promise of reward is high enough.

13.7 THE CROSS-LICENSING COP

Some industries have IP spaces that are so dense that it becomes necessary to have a collection of patents available for use in cross licensing, to avoid having either to take a license or to engage in litigation. One such industry is the semiconductor device industry. Since the invention of semiconductor devices in the 1960s, there has been a half-century of innovation and development in the basic semiconductor device components (transistors, junctions, capacitors, etc.) as well as innumerable semiconductor devices in the form of chips and circuitry that are used in everything from doorknobs to smartphones and computers.

The recent purchases by companies such as Google of large tracts of the semiconductor device IP space are not-so-subtle attempts to build an arsenal of patents that might be used to counter the threat of a patent lawsuit by others who work and live in the same IP space.

The COP for the cross-licensing strategy is

> **We want to have enough patents to be able to cross license our way out of trouble.**

This COP requires a relatively large patenting budget, a low-entropy patenting system, and a relatively low internal threshold for deciding which inventions to try to patent. A company that might be said to employ this COP is Applied Materials, located in Santa Clara, California. Applied Materials makes equipment, provides services, and develops software for manufacturing advanced semiconductor devices, flat panel displays, and solar photovoltaic products. Applied Materials has a market capitalization of about \$14.5B. It patents an extremely wide variety of inventions, from small-scale improvements to systems and methods to large-scale semiconductor manufacturing equipment.

Consider, for example, two Applied Materials patents. The first is US Patent No. 8,226,540, entitled "Configuration-based engineering data collection (EDC) for manufacturing lines," and has, as the title suggests, the scale of a semiconductor manufacturing line—which in case you do not know, is enormous. The second is US Patent No. 8,256,754, entitled "Lift pin for substrate processing," and covers, as its title suggests, a lift pin for lifting a substrate (wafer) from a substrate support surface—which in case you don't know, is one of the smallest components of a semiconductor-manufacturing tool. So, quite literally, Applied Materials seeks patents on the smallest to largest aspects of semiconductor manufacturing. Applied Materials receives, on average, about one patent a day from the USPTO and has thousands of patents.

As we discussed previously, having a large portfolio of patents allows the portfolio owner to take advantage of favorable patent statistics. Many of the patents will collapse to a valid state if they are measured. As we now know, a large portfolio gives rise to truly valid and valuable patents as well as patents that are probably invalid but are nevertheless valuable because of patent portfolio thermodynamics, as we discussed back in Chapter 4.

While the Applied Materials patenting effort is clearly not just for cross licensing, the wide spectrum of its patent portfolio makes the portfolio very useful for cross licensing if a patent dispute arises. Cross licensing is, in fact, the typical way patent disputes are settled in the semiconductor industry. You don't hear much about settlements like these because they tend to be made without much fuss, in contrast to the knock-down, drag-out fights that Rambus seems to enjoy.

IBM is another company whose business remains closely tied to the semiconductor industry despite its expansion in recent years into consulting services and technical support. IBM usually wins the "most US patents received in a year" prize, logging about thirty US patents a day and heaven only knows how many non-US patents. IBM collects billions of dollars in royalties from its patents and is generally considered a friendly licensor. It also uses its huge patent portfolio very effectively to cross license its way out of trouble. Of course, it also has the financial resources just to buy a company with patents it might like to have.

It is worth mentioning again how the cross-licensing strategy works for a technology business when confronted by an assertive licensor like Rambus. The short answer is: It doesn't. If Rambus approaches a semiconductor company seeking a license and all the wonderful royalty payments it would generate, it is not going to settle for a cross license. As we noted before, Rambus is not actually in the business of manufacturing anything, so it has essentially no use for cross licenses. More specifically, it is problematic to try to corner Rambus as an infringer of anyone's patents because it does not make, use, sell, offer for sale, or import any tangible goods.

13.8 THE ONE PRODUCT, ONE PATENT COP

Some technology businesses, in particular small companies and independent inventors, adopt a very simple and focused approach to patenting that can be summed up by the following COP:

> **We patent only those inventions that are embodied in a product, with each product being covered by a single, focused defensive patent.**

This is a decidedly defensive strategy, and it is not about going out looking for trouble. It is the antithesis of the assertive licensor's COP. Here, the patent owner is actually making a real product and is selling it or having it manufactured by someone else who sells it. In the latter case, a license is not used offensively but rather as an easier alternative to having to do the manufacturing and selling, which can sometimes be problematic for a small technology business.

Consider the small technology business Caldwell Photographic, which, as we discussed, is solely owned and operated by the independent inventor Brian Caldwell. One of Brian's COPs involves inventing specialized products for niche photography applications and then either having the product manufactured and sold by his company or licensing the patent to another company that is willing to manufacture and sell the product.

An example product that Brian and his co-inventor Wilfried Bittner invented is an optical attachment for reducing the focal length of an objective lens. The patent application covering this invention is cleverly entitled "Optical attachment for reducing the focal length of an objective lens."

Figure 13.5 shows the cover page of the published patent application.

While such optical attachments are not new, existing ones tend to have problems that perhaps only die-hard photographers can appreciate. The rear type of attachment that Caldwell and Bittner invented also overcomes a number of problems with existing attachments and allows for certain type of lenses called "single-reflex lenses" that are used with a camera having a mirror (so that you can look through the lens) to be used with a camera that does not have a mirror.

The point here is that the focal-reducer product that Caldwell Photographic sells is exactly the invention that is described in the Caldwell–Bittner patent application. This stripped-down patenting approach is suitable for a small technology business that has a single product in a niche market. Because the two inventors work and live in this and other niche markets of the photography and cinematography industries, there is a high probability that the product will be reasonably successful and therefore justify the patenting effort and costs.

A key component of the patent strategy for the one-product, one-patent COP is that the patent application be really good. It can't be a piece of junk thrown together on the cheap. "Costume jewelry" types of patents don't work well here. The patent

US 20130064532A1

(19) **United States**

(12) **Patent Application Publication**   (10) **Pub. No.: US 2013/0064532 A1**
   Caldwell et al.   (43) **Pub. Date: Mar. 14, 2013**

(54) **OPTICAL ATTACHMENT FOR REDUCING THE FOCAL LENGTH OF AN OBJECTIVE LENS**

(76) Inventors: **J. Brian Caldwell**, Petersburg, VA (US); **Wilfried Bittner**, Tsuen Wan (HK)

(21) Appl. No.: **13/589,880**

(22) Filed: **Aug. 20, 2012**

Related U.S. Application Data

(60) Provisional application No. 61/573,847, filed on Sep. 13, 2011.

Publication Classification

(51) **Int. Cl.**
    G02B 9/34   (2006.01)
    G03B 19/02   (2006.01)
(52) **U.S. Cl.**
    USPC **396/71**; 359/781

(57)   **ABSTRACT**

An optical attachment configured to be operably attached to the image side of an objective lens to reduce the focal length and focal ratio of the objective lens. The focal-reducing attachment includes four lens elements and has a magnification of between 0.5 and 1. The focal-reducing lens can work with objective lenses having relatively large working distances for a large format size as well as with cameras having a smaller format size and relatively small permissible working distance.

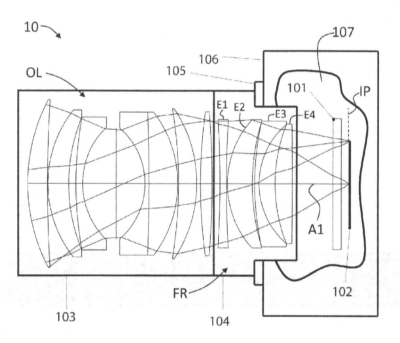

Figure 13.5 Cover page of the Caldwell and Bittner patent application.

quality has to be high. This means that serious work is needed in fleshing out the invention, identifying its unique aspects, and making sure that it is adequately described, shown in the figures, and claimed. A patent attorney with the right technical expertise needs to draft the patent application, and the inventors need to review the document carefully for technical accuracy and to ensure that the claims properly covered the commercial product. Since the patent will have to stand alone when challenging an infringer, its patent wavefunction needs to have a high probability of collapsing to a valid state if a measurement is forced upon it.

To this end, it is helpful for the inventor to have both specialized technical knowledge in the field and some business savvy about the relevant markets. This may sound obvious, but it is not always the case. There always will be people that try to patent niche inventions in fields far outside their expertise. It is worth remembering that most IP spaces are already crowded with inventions by people who have been there and done that. The art of photography, for example—the field in which the Caldwell–Bittner focal reducer operates—has been around for almost two centuries. During that time, a lot of smart people have invented countless types of photographic lenses and gadgets, including adapter lenses. There isn't much room for clueless people to wander into this kind of IP space and reinvent the prior art. Technical expertise combined with market knowledge and an appreciation for patents and the patenting process can go a long way toward ensuring that both the patent and the product it protects are solid and serve the intended business goals.

13.8.1 Large companies can effectively employ this COP

Small companies are not the only ones that seek single patents covering a single product. Most large technology businesses do so, but they also tend to seek patents on lots of other inventions that are not covered by their products, making the single-patent, single-product COP just one of several COPs.

A variation of the single-patent single-product COP reads as follows:

> **Some of our patents are defensive and are directed to covering a single product that we make and sell or intend to make and sell.**

Examples of this COP in action are numerous in the extreme. To find one, simply examine any product that is for sale and see if it (or its package) is labeled with a US patent. If the patent is owned by the same company that sells the product, then you are looking at an example of this COP.

13.9 PATENTING PRODUCT COMPONENTS COP

Another defense-based COP is directed to protecting components of a product. This COP reads as follows:

> **We patent the components of the products we sell.**

This COP can be seen in action in any industry that uses fairly complex machines. One such industry is the semiconductor tool industry, where the tools can be quite large and intricate. An example of such a tool is a photolithography tool that illuminates masks having extremely fine circuit features and images them onto a semiconductor wafer as part of the process of making integrated circuits.

Some of these tools are called "steppers" because the wafer is stepped and exposed in sequential fashion to create a number of identical exposure fields on the semiconductor wafer. Because the circuit features being imaged are amazingly small (much smaller than a micron, and hundreds of times smaller than the width of a hair), the lenses used in these machines are incredibly complex and expensive and are very difficult to manufacture.

One company that makes steppers is Nikon Corporation, based in Japan. Because the stepper's lens is the heart of that machine, it is usually also the subject of a patent. An example stepper-lens patent is Nikon's US Patent No. 6,700,645. The lens diagram from the '645 patent is reproduced in Figure 13.6.

Take special note of the complexity of the lens and the huge number of lens elements. What is not obvious from the lens diagram is that each of the lens surfaces needs to be finished to the highest possible accuracy and the lens needs to be assembled to extremely high tolerances.

Claim 1 of the Nikon patent is fairly lengthy though, relative to the complexity of the disclosed lens, it is also fairly broad. Yet, the '645 patent is fairly characterized as a "zone 4" defensive patent. Nikon uses this lens in one of its steppers and it is highly unlikely that the company is going to license it to anyone. It simply wants to ensure that it has the right to use the lens and that no one else tries to patent the same lens.

How likely is it that Nikon would ever need to sue someone for infringing the '645 patent? The short answer is: highly unlikely. The truth is that, even though the recipe for making the lens is right in the '645 patent, the lens is so specialized and complex and is so hard to get to work that most companies would lose their minds trying to make it, never mind trying to find someone dumb enough to buy it without the name "Nikon" on the lens barrel.

13.9.1 The automotive industry and patenting components

The automotive industry also patents components of its larger systems. In a recent television commercial, Mercedes-Benz boasts that it has 80,000 patents covering its automobiles. To drive the point home, the commercial shows one of its cars blazing down a coastal road trailing pages of patents behind it like so many leaves (antilittering signs are conspicuously absent). While the 80,000 number is a bit hard to believe, suffice it say that the major car companies have patents on major and minor components of their automobiles.

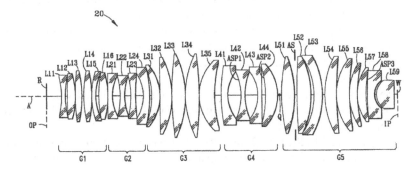

Figure 13.6 Lens diagram from US Patent No. 6,700,645.

13.9.2 Patenting every possible component: Apple and the iPhone®

Another variation of the patent-the-components COP is to patent every conceivable aspect of a product. This COP reads quite simply as

> **Patent every possible aspect of the product.**

This is what Apple sought to do with its iPhone mobile digital device (i.e., smartphone) and almost quotes verbatim Steve Jobs's directive to his team about what he wanted for patent coverage. Smartphones and smartpads have, despite their small size, a large number of components and features, from the cover glass and touchscreen down to the chips and software that control how the devices operate (e.g., scrolling, tap-to-zoom, bounce-back scrolling, etc.). These components and features are all the subjects of patents. The fact that the patents are owned by a variety of companies makes the smartphone business a tangled mess.

Apple happens to own a fair number of the key component patents for smartphones and smartpads. The widely publicized patent battle between Apple and Samsung in 2012 ended in Samsung being found liable to the tune of about $1B for patent infringement of some of Apple's utility and design patents on its iPhone and iPad technology (the judgment was later cut to about $600M).

At the time of this writing, an estimated fifty patent-related lawsuits pertaining to smartphone technology are under way. The smartphone market is estimated at about $220B.

NOTES

1. The interested reader is directed to Bouwmeester, D., A. K. Ekert, and A. Zeilinger, eds. 2000. *The physics of information processing: Quantum cryptography, quantum teleportation and quantum computation,* chapters 1 and 2. New York: Springer–Verlag. This is an excellent if not somewhat mind-bending discussion of how quantum cryptography works.
2. Shedeene, Jesse. 2010. The history of 3D movie tech. http://www.ign.com/articles/2010/04/23/the-history-of-3d-movie-tech
3. An interesting tidbit about this patent is that one of the inventors, Lenny Lipton, cowrote the song "Puff the Magic Dragon" back in 1963 with Peter Yarrow of Peter, Paul and Mary fame.
4. 35 USC §200-212 and 37 CFR §401.
5. See *Univ. of Rochester v. G. D. Searle & Co.,* 358 F.3d 916 (Fed. Cir. 2004).
6. See Hindle, Alistair. 2005. Reach-through claims. http://www.hindlelowther.com/article_reach.htm
7. See, for example, Riordan, Teresa. Patents; licensing boutiques help inventors with patent claims against big companies. *New York Times,* June 10, 2002.

GLOSSARY OF USEFUL TERMS AND ABBREVIATIONS

America Invents Act (AIA): Formally called the Leahy–Smith America Invents Act, the AIA was signed into law on September 16, 2011 (the main provisions became effective on March 16, 2013). It brought about a number of key changes to the US patent laws, including changing the US patent system from a first-to-invent system to a first-to-file system. The basic changes include an expanded prior user rights defense, the elimination of tax strategy inventions, new post-grant proceedings, changes in what constitutes "prior art," and a revised fee structure that now includes a new microentity applicant status that qualifies for reduced USPTO fees.

assertive licensor: A patent owner (or his or her representative) who aggressively seeks to compel others to license the patent, usually with a direct or thinly veiled threat of litigation. Also **offensive licensor**.

Bayh–Dole Act: A law that allows inventors associated with nonprofits, small businesses, or universities to retain patent rights to inventions made with federal funding. The law is set forth in 35 USC §202-2012 and the regulations are set forth in 37 CFR §401.

bottleneck patent: As the name implies, others must pass through (i.e., make use of) a bottleneck patent to access a particular technology. Also **pioneering patent.**

bureaucratic quantum tunneling: Akin to the phenomenon of quantum tunneling, whereby a particle can penetrate a barrier that is impenetrable according to classical mechanics, bureaucratic quantum tunneling permits a patent to issue from the USPTO that, by all indications, is invalid and that under classical patent mechanics should not have issued.

business–legal–technical (BLT) balance: A balance that considers the business, legal, and technical aspects of the patenting process or a given patent-related issue in equal proportions so that no one viewpoint dominates.

canonical patenting system: The most fundamental view of a patenting system, described by five elemental functions relating to innovations: generate, document, review, protect, and leverage. All functions are informed by the **central organizing principles** that mandate why a particular patent is being pursued.

central organizing principle (COP): The fundamental reason behind an activity or action. In the context of patents, the COP extracts the fundamental reasons why patents are being pursued.

claim: A statement in a patent application or patent at the end of a document that defines the scope (in legalese, the "metes and bounds") of the invention. There are two main types of claims: independent claims, which stand alone, and dependent claims, which refer back to and further limit a preceding claim. Limitations set forth in the specification are not supposed to be read into the claims.

claim construction: The act of construing one or more terms in a claim, where the meaning of the one or more terms is not entirely clear or otherwise subject to interpretation. See **Markman hearing.**

claim distortion: The alteration of a claim's scope when a patent application or patent is subject to processing by the USPTO or a court of law.

claim template reconstruction: When a patent examiner fails to read and understand a given claim in its entirety ("as a whole") and improperly uses the claim as a template to piece together the different claim elements from different prior-art references.

classical patent mechanics: A view of patents and patenting that relies on deterministic outcomes and linear thinking. Like the classical mechanics view of the physical world, classical patent mechanics only approximates the characteristics of patents and the patenting process.

client/law firm IP disconnect: When a client and the client's law firm each make incorrect assumptions about what the other is thinking due to miscommunication or the absence of communication.

Code of Federal Regulations (CFR): A codification of the general and permanent rules published in the Federal Register by the executive departments and agencies of the federal government. The regulations (i.e., the administrative law) regarding patents are set forth in Title 37 of the CFR.

coefficient of shame (C_s): A coefficient that ranges from 0 to 1 and that subjectively measures the amount of shame associated with submitting an innovation disclosure or filing a patent application on a given innovation, with 0 corresponding to having absolutely no shame (i.e., willing to submit or file on anything and everything, without any regard to prior art) and 1 corresponding to infinite pride (i.e., unwilling to submit or file anything, even if infinitely patentable).

commercial implementation patents: Patents that cover inventions directed to commercial embodiments (products) of a technology. Also called **commercialization patents.**

commercialization patents: See **commercial implementation patents.**

confidential disclosure agreement (CDA): See **nondisclosure agreement (NDA).**

confidentiality agreement: See **nondisclosure agreement (NDA).**

copyright: A form of legal protection for "original works of authorship," including literary, dramatic, musical, artistic, and certain other intellectual works. Copyrights generally protect forms of expression.

core invention: First invention in an IP space. See **pioneering invention.**

Court of Appeals for the Federal Circuit (CAFC): The highest federal court below the Supreme Court that hears appeals of patent cases from federal district courts and in some cases directly from the USPTO.

defensive patent: A patent that seeks to protect a product or process in a relatively conservative manner using tightly drafted claims that are not intended to initiate litigation but that could still be used to pursue an infringer who copied the product or process directly.

designing around: Evaluating a patent claim and coming up with ways to adjust or modify a product or process to avoid infringing the claim.

design patent: Patent directed to the ornamental (i.e., nonfunctional) aspects of an article of manufacture, such as containers, cases, lamps, computer icons, jewelry, etc.

doctrine of equivalents: A legal doctrine relating to claims and infringement, wherein claims are expanded to cover an equivalent of a claimed feature when the equivalent has substantially the same function and performs the function in substantially the same way to achieve substantially the same result.

dynamic range of operational details: A subjective scale that measures the amount of detailed thinking required for a given type of business operation, expressed in terms of frequency, wherein the scale ranges from the highly detailed thinking (high frequency) to the big-picture thinking (low frequency). When the business operation is a patenting system, the dynamic range encompasses low-, mid-, and high-frequency zones that are respectively associated with the business, technical, and legal aspects of working with the patenting system of a technology business. The concept is used to understand and explain why a balance between the business, technical, and legal aspects of the patenting system is required for low-entropy operation.

entropy: The amount of disorder in a system. Entropy is measured as a function of the number of ways that constitute "order" versus the system's total number of possible states. In the context of a technology business's patenting system, it refers to a measure of how well organized the patenting system is as a function of following or not following best-practice procedures versus all the possible ways of doing things wrong or inefficiently.

ex parte: One side or one party, such as in the patent examination process that involves one party (the applicant) and the examiner, who is not another party but is more like a minijudge. See also **inter partes.**

file history: A written record of the exchanges between the patent applicant and the USPTO concerning a patent application or issued patent. The file history includes things such as office actions, office-action replies, the filing of an information-disclosure statement, notices regarding informalities in the application, and replies from the applicant.

first to file: A law determining that when multiple patent applications on the same invention are filed in the patent office, the first inventor to file a patent application will be the inventor entitled to pursue the patent. The first-filed patent application will serve as prior art to subsequently filed patent applications on the same invention, thereby precluding the subsequent filers from pursuing a patent. The United States instituted as part of the America Invents Act (AIA) a modified first-to-file law, which became effective on March 16, 2013. The reason the first-to-file law is "modified" is because it allows for a 1-year grace period for the inventor for certain disclosures of the invention by the inventor or someone who obtained the subject matter directly or indirectly from the inventor.

fractal nature of innovation: A characteristic property of innovations whereby innovations beget innovations, leading to a cascade of innovations at finer and finer scales in a manner that resembles a self-similar fractal pattern.

freedom to operate: The ability to make, use, sell, offer to sell or import products, or to practice a method or process without risk of infringing the patent of another.

freedom-to-operate opinion: A legal opinion based on an analysis of the pertinent prior art with respect to a particular product or method to be made, used, sold, offered for sale, or imported into the United States that addresses the question of whether there is freedom to operate.

improvement patent: An invention that improves on an existing invention to the point where the improvement is patentable. Practicing an improvement patent may result in infringing a patent on the existing invention because patentability and infringement are two separate concepts.

independent inventor: An inventor who works mainly, though not necessarily exclusively, by himself or herself when inventing.

information disclosure statement (IDS): A listing of prior-art references submitted to the USPTO in connection with the duty, on the part of the applicant and those involved in the patent application process, to disclose information material to the examination of the patent application.

infringeability: A measure of a patent's ability to be infringed that to first approximation depends on the breadth of the patent's independent claim(s).

injunction: A court order directed to either stopping one party from taking certain actions or compelling a party to take certain actions.

innovation disclosure: A document used to record an innovation. Also **invention disclosure** or **record of invention.**

innovation quenching: See **invention quenching.**

intellectual property (IP): A catchall phrase that encompasses forms of intangible property and includes **patents, copyrights, trademarks, trade secrets,** and **know-how.**

inter partes: "Between the parties." Two sides or two parties in a legal action, which makes the legal action adversarial as compared to a one-sided or **ex parte** legal action. See **ex parte.**

invention: An idea, concept, device, process, and so on that is enabled—that is, that one can explain how to make and use. The transporter device on Star Trek is not an invention because it is not enabled; that is, no one knows how to make it and use it.

invention award/reward system: A business tool that gives cash awards based on a person's participation in the patenting system. The cash award is usually triggered by an event such as the filing of a patent application or the issuance of a patent on which the person receiving the reward is an inventor.

invention disclosure: See **innovation disclosure.**

invention quenching: When a person makes a unilateral decision not to document and submit his/her innovation or invention to the technology business at which they work.

inventor: A person who contributes to at least one claim of a patent application. Inventorship can change as claims change.

IP best practices: Those processes, procedures, and activities that are generally recognized as being the most efficient in connection with IP matters in general and patenting matters in particular and that are known to minimize patent system **entropy** when properly implemented.

IP-BITT: An IP-related statement that starts with the phrase "But I thought that…" and is either entirely incorrect or inaccurate.

IP performance metric: A managerial tool used to tie compensation to performance related to whether established IP-related goals have been met.

IP space: The patented and unpatented (i.e., claimed and unclaimed) aspects of a particular technology. The unclaimed aspects are called "green space." An IP space is considered densely packed when there are many patents with claims that cover very similar aspects of the technology—that is, when there is very little green space.

IP space black hole: An IP space that is so densely packed with patents that the repulsive force established by the obviousness requirement to keep at least some of the space open is eliminated. This renders ineffective the forbidden regions where inventions ideally cannot be patented so that essentially every invention in the space is patented.

IP zanshin: A concept based in the Japanese word *zanshin,* which in martial arts speaks to the concept of having such total awareness and emptiness of mind that you can act without thinking. In the context of IP, it means not having to think about taking actions consistent with IP best practices.

know-how: Practical information or institutional knowledge, often based on particular skills and experience, and usually directed to the (often undocumented) details about how to implement or practice systems or methods (including those related to the manufacturing of products or the delivery of services) for a given technology.

Leahy–Smith America Invents Act: See **America Invents Act.**

legal opinion: An attorney's formal legal evaluation of a given set of facts, a situation, or a particular legal issue.

long-form no-shame claim: See **no-shame claim.**

Manual of Patent Examining Procedure **(MPEP):** The manual used by patent examiners at the USPTO to examine patent applications. The MPEP's two printed volumes are collectively 4 inches thick and weigh about 10 pounds.

Markman hearing: A hearing that occurs in the course of patent litigation in a US district court, wherein the judge hears arguments and then makes a ruling on how certain words and phrases in a claim that are subject to different interpretations by the parties are to be construed. It is also called a **claim construction hearing.**

nondisclosure agreement (NDA): A contract that obligates the parties to maintain certain information as confidential and that can place limits on the use of that confidential information. Also **confidentiality agreements** or **confidential disclosure agreements (CDAs).**

nonpracticing entity (NPE): A patent owner who does not practice the patented invention. When such an owner seeks to enforce the patent, he or she is often referred to pejoratively as a **patent troll.**

nonprovisional patent application: A patent application that is examined by the USPTO and that may claim priority from an earlier filed application.

no-shame claim: A claim that blatantly covers the prior art. The no-shame claim takes one of two basic forms: the **short-form no-shame claim** and the **long-form no-shame claim.** The former is terse and easy to spot, while the latter is camouflaged with superfluous patentese and can be more difficult to spot.

offensive licensor: See **assertive licensor.**

offensive patent: A patent used for licensing or litigation that generally includes claims drafted in a broad manner so as to capture a large number of infringers. The claims may also be drafted in a focused way so as to capture a select potential infringer.

patent: A time-limited monopoly on an invention in exchange for an enabling disclosure of the invention. The time limit is generally 20 years from the earliest priority date, with certain exceptions.

patentable invention: An invention that meets the requirements for a patent as set forth in Title 35 of the **United States Code (USC).**

patent agent: A nonlawyer that is admitted to practice before the USPTO and that can prosecute patents on behalf of patent applicants. Patent agents cannot render legal opinions or practice law.

patent application: A document filed with the patent office that discloses an invention and is examined by a patent examiner as part of the patent-issuing process.

patent assignment: An agreement that conveys ownership of a patent. The party assigning the patent is the *assignor* and the party receiving the patent is the *assignee.*

patent attorney: A lawyer that is admitted to practice before the USPTO.

Patent Cooperation Treaty (PCT): A treaty signed by 117 countries that entitles applicants to file one international application in a standardized format at an established receiving office and to have that application acknowledged as a regular national or regional filing in any state or region that is party to the PCT. The receiving office conducts a search and examines the PCT application. A PCT patent application does not issue as a patent; it serves as a placeholder for filing international applications in individual countries and regions (e.g., Europe).

patentese: The turgid language used in patent applications to describe inventions and in particular to draft patent claims. It is a form of legalese peculiar to patents.

patent evaluation: The analysis of a patent from a business, legal, and/or technical viewpoint; not to be confused with a **patent valuation.**

patent infringement: The unauthorized making, using, selling, offering to sell, or importing of a patented invention.

patenting around: Obtaining patents on inventions that surround a claim or claims of an existing patent. Usually done with improvement patents that surround a core patent and commercialization patents that surround improvement patents.

patent invalidity: Perhaps better called "claim invalidity," it refers to when one or more of the claims of a patent are found to be invalid (i.e., fail to satisfy one or more of the requirements that make for a valid claim).

patent liaison: A person who interacts with innovators to facilitate their participation in the patenting system and who performs select activities such as interviewing innovators and drafting innovation disclosures, assisting in processing legal documents (e.g., obtaining properly executed signatures), reviewing patent applications, and the like.

patent license: An agreement that allows for the use of a patent in exchange for a fee. It is akin to a real-estate lease.

patent map: A visual representation of an IP space based on one or more aspects of patents within the IP space (e.g., broadest claim scope, inventors, subject matter, select claim features, numbers of claims, etc.).

patent pending: A phrase that indicates that a patent is still being examined by the USPTO and has not yet issued.

patent portfolio: A collection of patents generally directed to a given technology and covering regions of the same general part of IP space.

patent portfolio thermodynamics: The rules governing patent portfolios when the portfolios are sufficiently large for the details of the individual patents (and patent applications) in the portfolio to be ignored and for the patents to be treated collectively. Akin to how thermodynamics is used to treat particle-based systems such as gases.

patent prosecution: The interaction of an applicant for a patent (or more often, the applicant's patent attorney or patent agent) with the patent office during the examination process and in some cases after a patent has issued.

patent quality: A measure of how well a patent meets all the patent requirements. It represents the patent's ability to withstand a challenge based on an assertion that the patent fails to meet all of the requirements.

patent requirements: Assuming a spherical patent, the main patent requirements are (in alphabetical order): enablement, inventorship, no inequitable conduct, nonobviousness, no sale, no public disclosure, novelty, priority date, statutory subject matter, utility, and written description. These requirement terms are defined in the text in Chapter 3.

patent strategy: A planned course of action and activities in furtherance of the one or more central organizing principles that define the fundamental reasons for pursuing patents.

patent system: Processes and procedures used in a business or organization to process innovations in an effort to file patent applications and obtain patents.

patent system inertia: A subjective measure of the resistance a patent system offers to its participants.

patent system project manager: A person who oversees and manages a patenting system and who bears the ultimate responsibility for its efficient operation.

patent troll: See **nonpracticing entity.**

patent validity: A finding by a court of law that a patent's claims (and in particular those claims that are being asserted to establish patent **infringement**) are legally binding. A patent is entitled to a "presumption of validity," which can be rebutted. This means that a patent's validity state is uncertain until it is measured by a court of law. Patent validity is perhaps better called "claim validity." See also **patent invalidity.**

patent valuation: An estimate of the monetary value of a patent.

patent wave: The arrival of patents into an IP space either all at once or over a period of time. The patent wave typically precedes the technology wave so that by the time a product shows up in the marketplace, the IP space for that product already contains patents.

patent wavefunction: A concept based on a quantum patent mechanics view of patents that describes how the patent can exist in a superposition of valid and invalid states until it is measured by a court of law. Akin to the wavefunction of quantum particles such as atoms, electrons, and so on. The patent wavefunction is said to "collapse" to either a valid or invalid state when the measurement is performed.

person of ordinary skill in the art (POSITA): A hypothetical or fictional person who works in a given field or "art" and who is assumed to represent the average level of expertise in the field. Used as a reference in measuring obviousness. A POSITA is assumed to have knowledge of the entire body of prior art, and in this sense is anything but "ordinary."

pioneering invention: An invention that defines a new technology such that it opens up a new IP space with little or no closely related prior art. A pioneering patent or core patent is the first patent in the new IP space. Also **core invention.**

pioneering patent: See **bottleneck patent and pioneering invention.**

presumption of validity: The legal presumption that an issued patent is valid and that puts the burden of proof of invalidity on the one alleging invalidity.

prima facie case: The evidence and arguments presented by one party that are sufficient to establish a position, proposition, or fact. In the case of a patent application and the question of a claim's obviousness, it is the evidence provided by the examiner that is sufficient to establish obviousness unless the applicant rebuts the evidence.

prior art: All the information related to an invention that predates the effective date of a patent or patent application on the invention.

product clearance: The evaluation and analysis of an IP space to determine whether a product would infringe any patents in that space.

provisional patent: A misnomer. There is no such thing as a "provisional patent," and you should avoid self-appointed experts who use this term.

provisional patent application: A patent application that is not examined and that serves to establish an effective filing date. The formatting requirements are less formal than those of a nonprovisional patent application, but the substantive legal requirements are the same.

quantum patent mechanics: A sophisticated view of patents and patenting that takes into account the many sources of uncertainty in the patenting process. This uncertainty gives patents stochastic properties akin to the properties of quantum particles in a quantum mechanics view of the world.

read on: A term used with reference to a claim of a patent or patent application, wherein if the claim were actually practiced, it would infringe the claim of another patent. There is no "infringement" when a patent issues with a claim that "reads on" the claim of another patent. See **infringement.**

record of invention: See **innovation disclosure.**

requirement filter: An evaluation by the USPTO of a given requirement in a manner that falls short of making a full-blown measurement of the requirement and thus at best tends to reshape the patent application such that it can pass through the USPTO.

secondary considerations: Additional factors that can be considered when assenting nonobviousness of a given claim based on evidence of commercial success, industry praise, unexpected results, copying, industry skepticism, licensing, and long-felt but unsolved need, wherein the evidence is in relation to the claimed invention and not to the invention as described in the body of the application.

service mark: Like a trademark, but for services. See **trademark.**

short-form no-shame claim: See **no-shame claim.**

specification: The part of a patent application or patent that describes the invention. The claims are usually considered to be separate from the specification.

spherical patent: A metaphor that reflects simplifying assumptions made about a patent, patents in general, or any patent-related issue that does not take into account a patent's internal details or content. Such simplifying assumptions can generate insight but can also lead to inaccurate or incorrect conclusions.

target licensee: Party that an offensive licensor identifies as a potential infringer and from whom the offensive licensor seeks a license, usually with an implied if not direct threat of litigation.

technology wave: The arrival into the marketplace of commercialized technology in the form of products either all at once or over a period of time.

trademark: Any word, name, symbol, device, or any combination thereof used or intended to be used to identify and distinguish the goods/services of one seller or provider from those of others and to indicate the source of the goods/services. Also **service mark.**

trade secret: Information that is not generally known to the public, that provides some business or other competitive advantage, and that is kept secret using reasonable efforts. Trade secrets are governed by state trade-secret law. Most states have adopted some form of the **Uniform Trade Secrets Act (UTSA).**

unenforceable: The inability to enforce any of the claims of a patent due to a finding of inequitable conduct during the process of obtaining the patent.

Uniform Trade Secrets Act (UTSA): A set of laws developed by the Uniform Law Commission (ULC) to provide a framework for the fifty states (and US territories) to establish improved trade secret laws, with the intent to have some uniformity in state trade secret laws, particularly with respect to remedies relating to the misappropriation of trade secrets. The UTSA has been enacted in one form or another by forty-seven states (exceptions are Massachusetts, New York, and Texas, which rely on their state's common law decisions).

United States Code (USC): The compilation of the statutory federal laws of the United States. Title 35 of the USC sets forth the patent laws.

United States Patent and Trademark Office (USPTO): The office within the Department of Commerce charged with examining and issuing patents and trademarks.

utility patent: Type of patent directed to inventions having "utility" (i.e., functional use) that is the most common type of patent. The proverbial patents on mousetraps are utility patents. See also **design patent.**

validity uncertainty principle: Fundamental property of a patent wherein its validity (or, more accurately, the validity of the claims) is uncertain until measured. In terms of **patent quantum mechanics,** the uncertainty arises because a patent exists in a superposition of a valid state and an invalid state until its validity is measured.

INDEX

Index

T - #0510 - 101024 - C0 - 234/156/15 - PB - 9781439888056 - Gloss Lamination